Information and Communication Technologies in Nigeria

Society and Politics in Africa

Yakubu Saaka
General Editor

Vol. 19

PETER LANG
New York • Washington, D.C./Baltimore • Bern
Frankfurt am Main • Berlin • Brussels • Vienna • Oxford

Patience Idaraesit Akpan-Obong

Information and Communication Technologies in Nigeria

Prospects and Challenges for Development

PETER LANG
New York • Washington, D.C./Baltimore • Bern
Frankfurt am Main • Berlin • Brussels • Vienna • Oxford

Library of Congress Cataloging-in-Publication Data
Akpan-Obong, Patience Idaraesit.
Information and communication technologies in Nigeria: prospects
and challenges for development / Patience Idaraesit Akpan-Obong.
p. cm. — (Society and politics in Africa; v. 19)
Includes bibliographical references and index.
1. Information technology—Nigeria. 2. Information technology—
Africa, Sub-Saharan. I. Title.
HC1055.Z9.I553   384.09669—dc22   2008049791
ISBN 978-1-4331-0310-0
ISSN 1083-3323

Bibliographic information published by **Die Deutsche Bibliothek**.
Die Deutsche Bibliothek lists this publication in the "Deutsche
Nationalbibliografie"; detailed bibliographic data is available
on the Internet at http://dnb.ddb.de/.

© 2009 Peter Lang Publishing, Inc., New York
29 Broadway, 18th floor, New York, NY 10006
www.peterlang.com

All rights reserved.
Reprint or reproduction, even partially, in all forms such as microfilm,
xerography, microfiche, microcard, and offset strictly prohibited.

Printed in the United States of America

• DEDICATION •

I dedicate this book to my maternal grandfather, Chief Jackson Akpan Ekpo-Ekerete (1916-1985), and my father, Chief Timothy Titus Akpan (1944-2007). This is the culmination of their dreams and sacrifices for me.

Table of Contents

List of Illustrations xi

List of Tables xiii

Foreword xv

Preface xix

Acknowledgments xxi

Chapter One. Introduction 1
   Earlier Links Between ICTs and Socioeconomic Development 3
   Information Communication Technologies: A Conceptual Analysis 6
      The Internet as a Key ICT 7
   ICTs and Development: Nigeria as a Case Study 10
   Research Methodology 11
   Notes 14

Chapter Two. Information and Communication Technologies
  and Theories of Development 15
   Introduction 15
   Theories of Development 16
   Modernization Theory of Development Communication 21
      Modernization Theory and the ICT-for-Development Discourse 23
   ICTs and Development: Emerging Theoretical Perspectives 24
   Defining the Information Society 26
   Theories of the Information Society 29

The Stage Approach 30
Scenario-Modeling 31
Toward a Theoretical Framework of ICTs and Development 34
Constructing an Alternative Framework 34
Conclusion 35
Notes 36

Chapter Three. Evolution of Nigerian Economic Development 37
Introduction 37
The Nigerian Development Dilemma 38
Evolution of Economic Planning in Nigeria 41
Colonial Development 42
A Model of Development Plan for Nigeria 43
Ten-Year Plan of Development and Welfare for Nigeria, 1946–1956 44
Criticisms 44
Post-independenceNational Development Plan, 1962–1968 45
Criticisms 45
Second National Development Plan, 1970–1974 46
Criticisms 47
The Third National Development Plan, 1975–1980 47
Criticisms 49
The Fourth National Development Plan, 1981–1985 49
Criticisms 50
The Structural Adjustment Program Years, 1986–1988 50
The Buhari Administration 51
The Babangida Years and the Structural Adjustment Program 53
The Fifth National Development Plan, 1988–1992 54
Criticisms 55
Post-Fifth National Development Plan 56
Abacha and Vision 2010 56
Nigerian Economic Policy: 1999 to the Present 57
Objectives and Instruments 58
ICTs and National Development Plans 59
Conclusion 60
Notes 62

Chapter Four. A Journey to the Future: The Policy Framework 65
Introduction 65
The National Policy on Telecommunications (NPT) 67
Implementation of the NPT 71

Table of Contents

    Interconnectivity and Infrastructure Sharing.................................. 74
    National Backbone 75
  The National Policy for Information Technology (NPIT) 75
    Actors 78
    Implementation of the NPIT 79
  Contributions of the Private Sector 81
    State and Private Sector Alliances 82
  Conclusion 83
  Notes 84

Chapter Five. All in a Day's Work: Diffusion and Usage in the Public Sector 85
  Introduction 85
  Patterns of ICT Usage: Who Uses What, When and How? 87
    The Presidency 88
    Ministry of Science and Technology 90
    Federal Ministry of Information and National Orientation 91
    Ministry of Communications 94
    Federal Ministry of Education 98
  ICT Usage by Policy Implementing Agencies 100
    The Nigerian Communications Commission (NCC) 100
    Nigerian Information Technology Development
      Agency (NITDA) 102
  Analysis of ICT Usage and Diffusion in the Public Sector 102
    Pervasiveness 102
    Sophistication of Use 103
    Sectoral Absorption 104
  Prospects for E-revolution 105
  Conclusion 107
  Notes 109

Chapter Six. Use IT: Patterns of Usage in the Societal Context 111
  Introduction 111
  Penetration of ICTs in Nigeria: The Number Question 113
  Awareness and Perception of ICTs 115
    Socially Constructed Definition of ICTs 117
  Patterns of ICT Usage 119
    Most Frequently Used ICTs and Intensity of Usage 119
    Purposes of ICT Usage by Respondents 123
    Points of Access 126

Cost of Access 129
　　　Sophistication of Use 129
　　Attitudes and Expectations About ICTs 131
　　　Common Themes 135
　　Emerging Issues: Awareness, Access, Affordability and Availability 137
　　　Awareness 138
　　　Access and Availability 138
　　　Affordability 139
　　Conclusion 140
　　Notes 142

Chapter Seven: Potholes on the Information Superhighway ... and Detours 143
　　Introduction 143
　　Institutional Framework 144
　　　State of the Infrastructure 147
　　　Infrastructure Detours 151
　　　Poverty and Illiteracy 153
　　Ideological and Cultural Framework 153
　　　'Official Secrets' Versus Open Information Society 155
　　　Monopoly on Knowledge 156
　　Ethnicity as Pothole 157
　　　Ethnic Detours 159
　　Conclusion 160
　　Notes 163

Chapter Eight: 'African Giant' Meets Africa: Information and Communication Technologies Beyond Borders 165
　　Introduction 165
　　The African Giant and its Neighbors 167
　　Key Factors in Nigeria's ICT Growth 172
　　Prospects and Challenges for Continental Leadership 177
　　Conclusion 186
　　Notes 188

Bibliography 189

Index 197

Illustrations

Figure 6-1. Awareness as Measured by Identification of ICTs 115
Figure 6-2. Awareness of ICTs, by City 117
Figure 6-3 Most Frequently Used ICTs by Gender 120
Figure 6-4 Most Important Socioeconomic Concerns (Percentage of Times Selected) 132
Figure 8-1. African Top 10 Internet Countries 182

Tables

Table 4-1. NITDA's IT Parks 79
Table 5-1. ICTs Used in the Presidency, Aso Rock 89
Table 5-2. ICTs Used in the Ministry of Science and Technology 91
Table 5-3. ICTs Used in the Federal Ministry of Information and National Orientation 93
Table 5-4. ICTs Used in the Ministry of Communications 96
Table 5-5. ICTs Used in the Federal Ministry of Education 99
Table 5-6. ICTs Used in the Nigerian Communications Commission. 101
Table 5-7. Diffusion of ICTs in Selected Government Offices in Abuja 104
Table 6-1. Level of ICT Penetration in Nigeria 114
Table 6-2. Purpose of ICT Usage 124
Table 6-3. Destination and Origin of Last ICT Activity 125
Table 6-4. Points of Last ICT Access 127
Table 6-5a. Expectations About ICTs (Percentage of Respondents, 2001) 133
Table 6-5b. Expectations About ICTs (Percentage of Respondents, 2007) 135
Table 8-1. Top Ten Mobile Cellular Subscribers in Africa 166

Foreword

"Knowledge is Power," goes the ancient adage. This is even more so in contemporary times. Knowledge empowers the possessor because ideas are the building blocks of any creative and productive society. The trans-generational migration from Stone Age through the industrial revolution to the so-called information age has shown the catenation of the quantum and quality of thoughts and ideas. Present knowledge is used to drill into the unknown to discover new knowledge. There is logic sense therefore in the saying that the *Stone Age did not end for lack of stones but for the discovery of new engineering materials.*

Knowledge is the new material in the current Information Age because not only is it power, but it is also the source of wealth. It has become a veritable factor of production, a necessary resource for any society that wishes to progress. In the new economy, information and communication technologies (ICTs) accelerate the ease of and access to knowledge. Information and knowledge therefore now translate into productive ventures endowing a person, a people and a nation with competitive advantage in any field of human endeavor locally and globally. Many developing countries have explored the possibilities inherent in this new platform of acquiring knowledge for power in the new economy; Nigeria is not an exception though policymakers were initially slow to get on board. As recent as 1999, former President Olusegun Obasanjo was still skeptical about the possibilities of ICTs. Incidentally, he became the biggest supporter of ICT development in the country and the sector expanded exponentially during his eight-year administration. I once told him at an ICT event that since the logic of the internet contrasts with that of dictatorship, his commitment to the growth of ICTs in Nigeria might very well be the proof of his democratic credentials.

At the risk of hyperbole, I would declare that a digital revolution has occurred in Nigeria. Before the licensing of four Digital Mobile Operators in 2001, Nigeria, a potentially great country with a population of 140 million, had less than 500,000 telephone lines (both fixed and mobile). Today, the total number of telephone lines is nearly 55 million. In less than ten years,

Nigeria's teledensity has increased to 38.09 per 100 inhabitants. This is a remarkable quantum leap. Several factors have contributed to this phenomenal growth. These include a sufficiently deregulated and highly liberalized telecommunication environment, formulation of enabling policies, creation of regulatory and coordinating institutions, and a legal framework. Above all, Nigerians were starved of telecommunication services. Before the revolution, it took about a year to get a mobile phone service with a fee of about a thousand US dollars and an intensive lobby. In fact, owning a mobile telephone set was considered elitist and a status symbol. Little wonder the deluge in demand when, at the onset of the revolution, the service became available and affordable. In large part, the avalanche in demand is a release of pent-up tension.

There has been similar growth in the diffusion and penetration levels of other ICTs in the country. But what does this mean for Nigeria? Is the presence of ICTs in the country an end by itself or can the technologies promote socioeconomic growth in the country? How much have ICTs already contributed to Nigeria's economic indices in the last nine years? Professor Patience Akpan-Obong explores these and other questions in her book, *Information and Communication Technologies in Nigeria: Prospects and Challenges for Development*. She presents a historical account of the development of ICTs in the country, the role of various actors and the factors that facilitated and/or hindered the process. She examines diffusion and usage in the public and private sectors as well as in a subset of the general society. The result demonstrates a steady growth at all levels. Also, there is ample evidence of gradual broadband ICT infrastructure penetration. This book vividly shows concerted efforts at automation of processes in the Nigerian civil service as well as in the private sector.

Information and Communication Technologies in Nigeria: Prospects and Challenges for Development also addresses concerns about the development of Nigerian content. The demand for local content ranges from software to hardware prototype development, assembly and manufacture to fabrication of physical infrastructures like telecommunication towers and masts, and base station equipment enclosures. The author highlights some of the factors that are likely to hinder optimum exploitation of the potentials of ICTs in Nigeria. One of these involves the manner in which the misdeeds of an infinitesimally small number of "419" scam fraudsters tarnish the image of a nation of hardworking, honest, dynamic and resilient people.

The book concludes with an examination of a 21st century role for Nigeria in Africa. The country has always placed its human and material resources at the disposal of needy African countries. Despite its own problems, Nigeria is a peacemaker, beacon of hope and dependable "big brother" in Africa. The information age presents more opportunities for Nigeria to continue to be relevant continentally especially by modeling good corporate governance, zero tolerance on corruption, functional infrastructures, visionary leadership and development of a democratic polity. However, Nigeria cannot rise to the challenges and demands of continental leadership if it does not carefully address its own internal problems. Professor Akpan-Obong seems to believe that ICTs can facilitate a re-engineering of a more politically active pan-African Nigeria. It is perhaps too early to draw conclusions on this. However, I commend the author's efforts in raising the prospects.

As well, she continuously engages with questions which unveil the facts and shows Nigeria's history of ICT development in the making. She stimulates and sustains interest in the theme of ICTs as tools for socioeconomic development while nudging Nigerians to take stock of their accomplishments and perhaps to say: *for once we got it right*. She whets the appetite of the foreign venture capitalists to rush to this investors' paradise. This book is a refreshing mix of technical scholarship, journalism, and cosmopolitanism with a pan-African anchorage. It makes an exciting reading especially for those of us who have been involved right from the beginning of the ICT revolution in Nigeria. In a way, we are witnesses to history and it is a delight to see it chronicled in this manner. It reminds us of where we were not too long ago, where we are today and the trajectory going forward.

The book will remain for some time a compendium and resource book on ICT development in Nigeria. It is an indispensable working guide in the development of future ICT policies and programs in Nigeria and elsewhere in Africa. To the researcher and student of ICTs in development, the economic development planner and the would-be investor in one of the fastest growing ICT markets in the world, I strongly recommend *Information and Communication Technologies in Nigeria: Prospects and Challenges for Development*.

In closing, I would like to recognize my fellow "pioneers" of the new frontier in Nigeria. These men and women put their political and intellectual capital at the disposal of their country in the development of ICTs in the country: His Royal Highness Eze Cletus Iromantu, pioneer CEO of the Nigerian Communications Commission (NCC); Engr. Ernest Ndukwe, current CEO of the NCC; Chief Johnson Asinugo, an NCC Consultant; Chief Chima Onyekwere, CEO of Linkserve, Nigeria's premier ISP; Prof. Gabriel O. Ajayi, pioneer director-general of NITDA in whose tenure the IT Policy was formulated; Mrs Ibukun Odusote, pioneer administrator of the Nigerian Top Level Domain, .ng; Engr. Lanre Ajayi, CEO of PINET and current president of the Nigeria Internet Group; Engr. Titi Omo-Ettu, CEO of Telecom Answers Associates and publisher of Cyberschuul; Mr. Sunday Afolayan, CEO of Skaanet, one of the pioneer ISPs; Mr. Chris Uwaje, CEO of Connect Technologies Ltd; Mr. Sonny Aragba-Akpore, head of the ICT Desk at the *Guardian* Newspapers; Engr. Etim Amana, MD of MIS Software Ltd.; and a host of others whose names are worthy of mention, but the limitation of this space does not permit.

Emmanuel Efiong Ekuwem, PhD, NPOM,
National President, Association of Telecommunication Companies of Nigeria, and Group MD/CEO, Teledom Group.

Lagos, Nigeria
October 2008

Preface

A Personal Odyssey

My introduction to the issues of development, or rather, underdevelopment, began in Nigeria during ten years of work as a journalist. I started out in 1984 in the exciting and exacting field of sports reporting where the issues were clearly defined: a football (soccer) team won, lost or drew a match. As a sports reporter in a state-government-owned newspaper, *The Nigerian Chronicle*, work revolved around the fortunes of the state team, Rovers Football Club of Calabar, then in top form in the premier division of the national league. It won its home matches, drew many away matches and recorded few loses. Three years later I moved to a bigger newspaper (*The Punch*) and a bigger city (Lagos) where life was more fast-paced and the distinctions between losses and wins were not so unambiguous. Assigned to the Woman's Desk the challenges of work were however not demanding. All I needed was to perfect my writing skills and entertain my female readers on "ten ways to get your man" or some such fare. After a year of this, I decided to give the women's pages some "bite" and began a series of in-depth articles on the participation of women in the emerging "Third Republic" politics of the country in the late 1980s. I also focused on issues of women in poverty in a style uncommon on many women pages of the time. Suddenly, my urban, educated middle-class female readers were confronted with photos of women literally eking out a living from pieces of coal in the abandoned coalmines of Jos rather than those of the latest fashions out of Paris and New York. This political incursion on the women's pages became the appropriate segue to the newly created political desk, which I coordinated as an assistant news editor. That thrust me centrally into politics and began a journey that has since defined my scholarship and interests.

Journalism in Nigeria provided an opportunity to travel to many parts of the country. In the process, I saw the many faces of Nigerian poverty. Contrary to widespread assumptions in different parts of the country, poverty has no religion, region or ethnicity. Nigeria's colonial heritage had structured a problematic regional distribution of power in the country. The mainly Islamic and Hausa/Fulani North seemingly wielded political power with the mostly Christian and multi-ethnic South having most of the economic power. Since political power in Nigeria usually translates as access to economic power, the view from the South was that the North was better off than the rest of the country. This was particularly troubling to many Southerners because Nigeria's major resource, oil, is extracted from their region and there was the sense that these resources were being utilized to the service of Northerners. The view from the North was that the South had everything.

The opportunity to travel around the country showed that poverty is widespread and affects the majority of Nigerians regardless of region or religion. Those who had access to economic wealth were few and their enclaves were located in different parts of the country. As I followed some of the Third Republic politicians on their campaign trail, I realized that while they claimed to have the solutions to problems of poverty and inequitable distribution of resources, they all saw access to political power as synonymous with personal economic enrichment. I did not have the answers either, and this led to a lot of frustrations as weekly, I tried in a weekly column and in news features, to articulate the "way forward."

This frustration was to re-surface during my first year in the PhD program at the University of Alberta. In much of the development literature that I had come across, there seemed a repetition of mostly modernization-driven strategies of development that had been tried in many countries but failed to result in economic growth. Clearly, there was a disjuncture between theoretical analysis and the realities of people living in the Developing World particularly in sub-Saharan Africa. The frustration and urgency mounted for me because while I have taken on Canadian citizenship and could be considered materially comfortable, the pictures of malnourished black children and women continually haunt me. I am constantly confronted by the question: how can Africa escape this crushing poverty? In the search for answers, I came across the debates in the field of information and communication technologies (ICTs) and their linkages with socioeconomic development. I decided to explore the ICT-for-development (ICT4D) discourse in the hope that it held the answers. That curiosity has over the years grown into a research agenda emerging from an intellectual fascination as well as a personal and political odyssey. The journey took me back to Nigeria where I sought to understand the ways in which the country and its policymakers were engaging with the new technologies and their claims for development. The outcome of that journey forms the substance of this book, though the search for solutions to the problem of poverty in Africa continues.

Acknowledgments

I am profoundly grateful to those who did the tedious work of reading through the draft chapters of this book. At the top of the list is my department chair and mentor, Dr. Nicholas Alozie, who offered extremely helpful comments. His critique improved the quality of the argument and analysis in the book. A friend, Mr. Isa Olayiwola Mohammed, also provided useful suggestions. His knowledge of statistics made a huge difference in the presentation of tables and figures in the book. Two other friends, Dr. Mary Jane Parmentier (my colleague and research partner) and Ms. Janet Barr, proofread early drafts of some of the chapters. Dr. John Mbaku of Weber State University provided feedback during the conceptualization stage. Of course, all errors are entirely mine.

I particularly acknowledge the contribution my friends, mentors and professors at the University of Alberta. Their advice and support were pivotal to the initial research that forms the basis of this book. I am therefore eternally indebted to Professors Yasmeen Abu-Laban, Tom Keating, Fred Judson and Ann McDougall.

My research assistant, Ms. Nefertiti Aquah, read the final drafts of each chapter. She also compiled the index and bibliography. Her perspective as a graduate student and a "non-IT person" ensured that the tone and language in the book are framed appropriately for a wide range of readers. Nefertiti played another, equally important role. As my oldest daughter, she helped on the home front in caring for her younger siblings during the days and nights when this project took over my duties to the family. Also on the home front, my sister, Dr. Mercy Ette, sacrificed her 2008 summer to travel from Huddersfield, England, to help out with the family so I could write with minimal distraction. Ms. Lessina Cline and her mother-in-law, Ms. Celina Cline, lovingly supported me with their prayers and friendship.

I thank my husband, Timothy Obong, for his emotional and spiritual support and encouragement while I was writing this book, and for every moment of my life. He is my number one fan and one-man cheering team! I also thank my younger children, Mfon-Abasi (MJ), Iniobong and Menyeneabasi (TJ) for their cooperation and patience when I was too immersed in the book to function fully as a mother.

My parents, Chief Timothy and Mrs. Dorothy Akpan, started the process that has so profoundly impacted my life and the eventual ability to write this book. They believed in the value of formal (Western) education in a society and at a time when it was more "profitable" to marry off daughters than "waste" money in sending them to school. Regrettably, my father is not here to read this book. He was shot dead at the age of 63 by hired assassins in his living room on the evening of Nov. 28, 2007. Only the first three chapters of the book had been written at the time. Less than four months later, my younger brother, Blaise, shot the night my father was killed, also died.

The last five chapters of this book were therefore written in tears and enormous pain. I was so overwhelmed with grief that I considered giving up on everything. I kept going only because I knew I would have dishonoured my father's memory if I did not move forward: he never quit on his projects, not even when the going got tough. Blaise, who died on March 14, 2008 at the age of 27, also taught me a lesson about living life to the fullest while one can. He surrounded himself with serenity and knew how to simply sit and be. Now that this book is finished, perhaps I too can do that!

I thank Ms. Caitlin Lavelle and Ms. Jackie Pavlovic at Peter Lang. Their understanding, patience and guidelines helped during the writing and production of this book. Caitlin kindly extended the submission deadline so I could recover from my personal losses relatively enough to continue with the book. Jackie patiently guided me through the production process. Her attention to details is phenomenal!

Finally, I thank members of the Nigerian Youth Service Corps in Ikeja, Lagos (2001 and 2007), Port Harcourt and Abuja (2001) who participated in the research. I also thank the NYSC officials who allowed me unmediated access to the NYSC members, as well as shared their office space with me. I am grateful to the federal civil servants in Abuja and some private-sector stakeholders in Abuja and Lagos who agreed to be interviewed. Dr. Emmanuel Ekuwem and Engr. Titi Omo-Ettu were particularly helpful during both phases of the research. Dr. Edwin Madunagu in Calabar and, my brother, Mr. Francis T. Titus-Akpan in Uyo, provided essential logistic support in 2001; as did Mrs. Mary Essien in Calabar, and Mr. and Mrs. Nsikak Essien in Lagos.

Thank you, everyone!

Patience Akpan-Obong, PhD
Queen Creek, Arizona
December, 2008

• CHAPTER ONE •

Introduction

The potential of information and communication technologies (ICTs) to drive socioeconomic growth in developing countries is perhaps the most significant discourse in development theory and practice in the last two decades. The discussion is structured around two significant imperatives: the urgency to achieve socioeconomic growth within countries, and the necessity to integrate into an increasingly networked global economy. The major assumptions are that the technologies of information and communication such as the internet, telephone and computers will help developing countries to leapfrog several stages of development thereby catching up with or at least closing the gap between them and the postindustrial world.

These assumptions constitute a fundamental aspect of the general discourse on globalization and the notion of an integrated and interdependent world (Keohane and Nye, 2003; Held et al, 2003). As Castells (1996) argues, a combination of the new ICTs and the processes of globalization has created a global network society where the mode of production is informational, thus replacing the industrial-age mode of production. According to him, countries that fail to integrate the technologies into their national economic strategies will be excluded from the global network society. Indeed, Castells links the underdevelopment of Africa to its dismal level of technological development.

While much of this discourse on the utility of ICTs as tools for socioeconomic development is framed in language that often implies a technological determinism, many developing countries have responded by incorporating these technologies in their core development strategies. This has been achieved through the formulation of polices and implementation of projects and programs that deliberately target ICTs as means to economic growth, and as ends by themselves. In particular, African countries have embraced ICTs as viable tools for socioeconomic development. Also, international development agencies such as the Canadian International Development Research Center (IDRC) have funded ICT-centered research and projects. Indeed, toward the end of the 1990s, the discourse drove the programmatic agenda of the World Bank (Knowledge for Development Program) and the G8 (Digital Opportunities Taskforce). It remains an integral part of the United Nations Millennium

Development Goals (MDGs). These ICT-centered programs and initiatives proceed from a conceptualization of the capacities of the technologies in promoting independently, or in concert with other variables, the development goals of organizations, countries or regions.

However, there is a paucity of critical studies demonstrating the linkages between ICT application and socioeconomic growth. While many evaluative studies have begun, there is a lack of consensus on the direct effects of the application of ICTs on socioeconomic growth in developing countries. This book is aimed at contributing to the emerging research in this area through a case study of the process of harnessing ICTs in pursuit of socioeconomic goals in a pivotal African country. It also seeks to provide a critical-structural perspective to the ICT for Development (ICT4D) discourse by exploring the conditions (or structures) and contexts that facilitate or hinder the achievement of ICT-centered development goals in emerging states such as Nigeria. The book achieves its objectives through a rare multidisciplinary approach accessible to a broad spectrum of readers. It examines the role of the state and its institutions and structures in the diffusion of ICTs for development. It examines state-society relations and the new forms of power relations and class formations that the technologies are likely to engender, especially in countries with high levels of income disparities.

The book achieves its purposes by addressing a fundamental question: Does the trajectory through which the ICT sector was developed in Nigeria position the country to effectively harness the technologies for socioeconomic development? Responding to this question necessarily leads to an examination of Nigeria's policies on ICTs, their applications as strategies for socioeconomic development in the country and the societal response. It also involves an analysis of the conditions that hinder or facilitate the achievement of ICT-intensive socioeconomic goals. The results from this process will enable one to reach some tentative conclusions regarding the outcomes of the prevailing theoretical and policy assumptions about ICTs. These conclusions potentially have implications beyond Nigeria's borders especially if the technologies enhance Nigeria's capacity to participate more actively in the socioeconomic and political processes in Africa.

Information and Communication Technologies in Nigeria: Prospects and Challenges for Development also acts as a compendium of the history of ICT development in Nigeria and concentrates on the period from 1992 to 2007 when several notable developments in the ICT sector occurred. These include the establishment of the Nigerian Communications Commission (NCC), a key actor in Nigeria's ICT industry, the deregulation of the telecommunications industry, the entrance of the private mobile telephone service providers, formulation and implementation of the policies and legal framework on telecommunications, and ICT. The focus on policy and the legal framework is underscored by the assumption that the role of the state in harnessing ICTs for development is pivotal to the achievement of Nigeria's overall socioeconomic goals. The state in Nigeria (as is in many other developing countries) is a key participant in the development of ICTs in the country. Its involvement especially through regulating the sector, has orchestrated growth in ICT usage in the country. For example, the licensing of the initial three operators to provide mobile telephony services using the Global System of Mobile Communications (GSM) technology led to a rapid increase in teledensity in the country.

Also, previous development agenda in Nigeria (and other post-colonial states) have always been pushed by the state, a consequence of the colonial legacy in which African states were born fully formed but without a strong civil society. This is particularly so given that for the greater part of Nigeria's post-independence history, it was governed by non-elected military regimes. As Jarmon (1988: 7) asserts, the

> prime mover behind the process of national development in Nigeria is the national government. For example, no other institution possesses a similar relation to power and the control of coercive force; no other controls the major source of economic production and resource allocation; no other determines national goals and the means of their implementation; and no other holds the diverse Nigerian population together. These roles have emerged with the state in Nigeria and continue to be of central significance.

Western market economists might argue that it is precisely the intervention of the state that accounts for Nigerian underdevelopment, but such arguments understate the role of the state in the development (or industrialization) process in the West. It is not unusual therefore that in the age of ICTs as a paradigm of development, states will be at the forefront of efforts to harness the technologies for economic development. Indeed, this has already occurred in Nigeria as the state adopts a unique ICT policy of simultaneously controlling and deregulating in ways that may account for the rapid growth of the sector.

Earlier Links Between ICTs and Socioeconomic Development

The conceptual linkage between ICTs and development is as old as development theory, an approach that emerged in the late 1950s and 1960s at the height of decolonization. This coincided with the ascendance of the United

States as the global hegemon following the end of the Second World War and the ensuing Cold War. As a global hegemon, the United States was interested in the fate of the newly decolonized and independent countries, mostly to discourage them from turning to the Soviet Union for assistance. There was therefore a flurry of policies to assist the new countries to industrialize, particularly along the capitalist model of the US. The development of telecommunication infrastructures was emphasized as a major factor in facilitating the rapid economic, social and political changes in these countries. Particularly, decolonized countries were encouraged to develop their media such as print and broadcasting. Beyond being indices of "modernization," these technologies and infrastructure were expected to expose the people to information that would aid in their processes of development. Consequently, in the 1950s the UNESCO established a threshold of access to media of communication.

For a country to seriously begin the process of development, the UN agency contended, it had to "provide ten newspaper copies, five radio receivers and two sets of cinema seats for every 100 inhabitants" (UNESCO, 1955). Exposure to these media was expected to automatically enable people in "traditional" societies to acquire skills and attitudes that would usher them into the dawn of a modern (western) era of development. The problem was that many policymakers in developing countries embarked on the wholesale importation of the technologies from the industrialized world without concentrated efforts at indigenous development.

During the oil boom era of the 1970s, Nigeria, a nation noted for importing everything from capital-intensive products to consumer items such as the common salt, awarded enormous contracts for the procurement of telecommunication equipments to develop the country's telecommunication infrastructure. One of such contracts, awarded to the International Telegraphs and Telecommunications (ITT), haunted the presidential campaign of late Moshood Abiola, presumed winner of the botched presidential elections of 1993. He was the executive vice-president for Africa and the Middle East at ITT when the contract was awarded. His political opponents accused him of self-enrichment by procuring used and obsolete equipment. Abiola always denied the allegation insisting that the imports were in accordance with contractual specifications. The problem, he argued, was inappropriate storage. In any case, they were never installed and many critics held the failure of that initiative responsible for the poor telecommunications infrastructure that plagued Nigeria over the years.

Nigeria was not the only country where the spirited effort to develop communication and information infrastructure was still born. Even in countries where adequate infrastructural frameworks were available, little progress occurred in part because critics were beginning to raise questions concerning the direction of information flow and the real beneficiaries of ICTs. These questions crept onto the agenda of the UNESCO and the United Nations General Assembly and resulted in a formal discussion about the need for a New Information and Communications Order. "When the dust subsided, many developing countries—adverse experiences notwithstanding—expressed a strong interest in receiving foreign aid to develop their information and communication infrastructures (Hamelink, 1997:13). The concern for the state of telecommunications in developing countries led to the establishment of the Maitland Commission at a conference of the International Telecommunication Union (ITU) in 1982.

Three years later, the Commission released its report, *The Missing Link*. The report indicated an urgency to prioritize investment in communications, improved effectiveness of existing systems, new methods of financing to deal with scarcity of foreign exchange in developing countries, and a more effective role in ICT development by the ITU (Hamelink, 1997). The report, "revealed major differences in the provision of basic telephone services throughout the world" (Credé and Mansell, 1998) and prompted renewed interest in ICTs as "technologies that could markedly improve industrial performance and increase economic activity. The notion of leapfrogging became prevalent as it was widely suggested that through ICTs, developing economies could rapidly advance to post industrialization. Hamelink points out that while developing countries saw the potentials that these technologies had for their socioeconomic growth, not many of them were eager to jump in because of concerns about the social risks.

> People were concerned about issues like the potential for cultural colonialism, the replacement of jobs by machines, and the erosion of individual privacy and national sovereignty. Towards the end of the 1980s these fears seemed to have abated, and the general view on the relation between ICTs and development entered a third and current phase (Hamelink, 1997: 14).

Similar issues are now embedded within the context of post-industrial societies. Scholars, such as Jurgen Habermas are known for their notion of the declining public sphere and integrity of information and Giddens' work on the surveillance state addresses questions of privacy.[1] Schiller (1986) was concerned about the threat to the sovereignty of some nation-states when other more powerful ones deploy their communication resources to obtain unauthorized information. Contemporary versions of these concerns include privacy of personal information, cyber crime and security, identity theft and pornography. Despite these concerns, debates about the inherent potentials that ICTs have for development have dominated the center stage of policy and theory.

Information and Communication Technologies: A Conceptual Analysis

In the literature, ICTs are sometimes simply referred to as information technology (IT) with the obvious emphasis on their functions for the storage and dissemination of information (Dearnley and Feather, 2001; Avgerou, 1998, Dordick and Wang, 1993, Schiller, 1984a, 1986). In other instances, they are referred to as information processing technologies, a terminology that highlights how the technologies facilitate other activities (such as cataloguing health records to enhance the delivery of healthcare services). Still others employ the concept of ICTs (and sometimes in the singular) to capture the integrated nature of these technologies, and thus their prospects for, or challenge to, use as tools for socioeconomic development (Heeks, 1999; Credé and Mansell, 1998; Howkins and Valantin, 1997). "Information age" is also frequently used to stress the new "information mode of production" and the rise of information as a "standard in the realm of exchange value, replacing gold and raw materials" (Rajaee, 2000: 65). The UNDP *Human Development Report* for 2001 uses "network age" to stress the connections between ICTs and the processes of globalization. The different terminologies imply different emphases and approaches to the understanding of the implications of these technologies for the 21st century. It is not by accident therefore that the issue has been evaluated and analyzed from various disciplinary fields, most frequently management information systems/science, information science, informatics and computer science.

In the attempt to avoid the difficulty of defining IT or ICT, some authors now refer to the different technologies individually. This however raises new issues as many people commonly collapse ICTs into a single concept – computers, telephone, cell phones or the internet. Regarding ICTs as just access to a telephone or the internet may highlight the communication capabilities of these technologies, but it leaves uncontested their perceived capacities as agents of development. Besides, many of the technologies—such as the internet/computer or internet/cell phone—are "bundled" other and decoupling becomes problematic. This book therefore addresses all the technologies that constitute ICTs with a critical focus on their applications as tools for socioeconomic development. It acknowledges the convergence of the technologies and their multi-functional and multi-purpose applications. This perspective facilitates a holistic analysis of the socioeconomic implications of the technologies for developing countries.

The Internet as a Key ICT

All ICTs feature in the debates that link them with development, but many of the references to their capacities in development are usually about the internet, and the related hard/software—telephone, computer, modem and the access to the internet itself, databases and all aspects of the World Wide Web, and the technologies that enable networking. This explains the emphasis in the literature on how ICTs are enablers of information which in turn translates into knowledge and creates socioeconomic development. The internet also gives rise to Castells's concept of the global network society (1996), Webster's "information society" (1995)[2], Rajaee's information age (2000) and Keohane and Nye's "information revolution" (1977)—all of which are concepts that feature in the ICT4D discourse. According to Poster (1999), the internet combines

> all previous communications systems, amplifying them in new ways and integrating them with digitized computer apparatuses. (It) also provides the innovation of combining the decentralization of the telephone network with television's capability of reaching large audiences. And it improves on each of these... (p.235).

While the story of the internet has been told and retold, it does bear some overview. Dearnley and Feather (2001) credit Paul Baran, an American electronics engineer, with "inventing" the internet. Baran, while working for the Rand Corporation, developed his ideas about networks which he had nurtured in another context. He eventually formulated the concept of "subdividing information into small packets, which could then be transmitted through telecommunications networks between computers" (Dearnley and Feather, 2001: 30). According to this account, initially, the telecommunications companies were "suspicious to the point of hostility" but the military was interested in its prospects for nuclear defense.

> Computer scientists (at the Advanced Research Projects Agency, ARPA) came across Baran's work, and commissioned him to look at its defense potential. In October 1969, working with the University of California at Los Angeles (UCLA), ARPANET successfully established a network initially linking four academic computing sites. (Ibid)

The United Kingdom and France later embarked on projects that would subsequently enable international networking of computers. But internet activity occurred mostly within the military and academia until about a decade later when the US National Science Foundation, "the sponsor of much of the science that had created ARPANET, agreed in 1979 to allow commercial exploitation of the network" (Dearnley and Feather, 2001: 31). Still, it took another decade for the internet to leave academia and explode into the public domain, with 1995 marking the watershed year of the technology when usage doubled. By June 2008, global internet usage had risen to nearly 1.5 billion people, with just 3.5% or a little over 50 million occurring in Africa (Internet World Statistics, 2008).

The internet was arguably one of the most powerful inventions of the 20th century, even as there is no consensus on its revolutionary nature. Mackay (in Held, 2000) argues that the telegraph was more significant in shrinking time and space than the internet has been. In fact, "the history of the telegraph suggests that there is nothing dramatically new about recent communication technologies and global communication, and cautions us regarding the more apocalyptic claims which are made about 'the information revolution'" (Mackay Hugh, 2000: 71). Dearnley and Feather (2001) agree that the internet (particularly its public domain, the World Wide Web) has radically altered "the way in which we live and work" (p. 34).[3] But they also note that the internet is "merely a system for communicating data. It is powerful in terms of its speed and capacity but conceptually no different from any other communication system" (Dearnley and Feather, 2001: 44). Whatever the case, the internet has undoubtedly enabled easier, faster and cheaper means of communication in ways never before imagined. As Poster (1999) points out, "the digital form of information on the internet provides the advantage of virtually costless copying, storing, editing, and distribution" (p.235).

Information as a factor of production: The internet and related communication technologies have led to what Harvey (1989) referred to as "spatio-temporal compression." Combined with advances in transport technology, this time-space compression created what some scholars refer to as a global network society (Castells, 1996) or communication revolution (Mowlana, 1997). The common assumption is that we have moved from a production and extraction economy to a knowledge-based economy where information has become a crucial factor of production. People and countries that desire to advance in the new economy must be information-rich. The internet and other technologies of communication enable access to this vital information. In the new economy, people and countries are presumably operating from a level playing field with equal access to the same information especially through the unregulated and uncontrolled internet. This is the basis of the ICT4D discourse. It compels developing countries to exploit information to achieve their socioeconomic development goals.

That information plays a crucial role in the present global economy is not in dispute, but the assumptions about its imperativeness for developing countries raise a number of questions. For instance, has its pervasiveness resulted in a new socioeconomic divide: the information rich and the information poor? How does a continent with 3.5% of global internet usage compete in the information society? Is all information including those produced by the culture industries of the West (particularly the United States) relevant to the particular needs of poor people in developing countries? As Heeks (1999) argues, data (the first stage in the production of information) is

> created within a particular context and retains embedded characteristics of that context: it contains what its creators do know and do feel is important and misses out what they do not know or do not feel is important; it reflects their political and economic beliefs; it reflects their culture.

If information is the new factor of production and commodity of value, it is conceivable that its production will follow the pattern of one earlier factor of production, namely capital. This then leads to a contestation of the notion that poor countries can effectively participate in the new economy as equals in ways that benefit them and their citizens. Given the present state of the technology, is it likely that people living in peripheral countries will remain mere passive consumers of information (and technologies) produced elsewhere?

Also within countries, former patterns of social relations and the class structure will potentially exclude people of particular socioeconomic classes, age and ethnicity from the process of utilization of ICTs for socioeconomic development. In Nigeria, new socioeconomic cleavages are beginning to emerge between those with access to the technologies and those without. The "modernizing elites" (or, culture elites, as Chirot, 1994, refers to them) are pushing an agenda that privileges the inevitability of ICTs and their utility as development tools. And yet, access to the technologies is circumscribed by income, age, gender and geographical location (with those in the urban centers having relatively greater access than the 65% who live in the rural areas). The rapid diffusion of cell phone usage in the country (from 100,000 in 2001 to 50 million in 2008 and 100% digital spread, according to ITU data) might imply equitable access to this particular technology. However, even in that, there are socioeconomic cleavages in the context of handset types (the more expensive the better) and the number of networks to which an individual subscribes. Also, as the research presented in this book shows, greater usage of the technologies for personal purpose—such as socializing with friends and family— outweighs business transactions that might accrue toward macro socioeconomic gains.

These issues inform the need for a critical analysis of the process through which ICTs can be harnessed to achieve socioeconomic goals in ways that do not exacerbate preexisting socioeconomic inequities in developing countries. Nigeria has been chosen as a case study to explore the process through which ICTs promote socioeconomic growth. The country makes an excellent because of its potential role in Africa. For one, it is the most populous on the continent and has the largest economy. The process through which the country harnesses ICTs for development could become a model of best practices—or loopholes to avoid—for other countries on the continent. One therefore hopes that the portrait painted in this book will facilitate understanding of the larger question of the connections between ICTs and socioeconomic development both in Nigeria and other African countries.

ICTs and Development: Nigeria as a Case Study

Nigeria is a rather late entrant to the debates on the relationship between ICTs and development. As noted by a major player in Nigeria's ICT sector, the years of military dictatorship in Nigeria coincided with the revolution in information technology (Ajayi, 2001).[4] For instance, the first known e-mail activity in Nigeria occurred only as recently as 1994, when many other countries in Africa were already ahead on the "information superhighway." With the formulation of two policy documents in 2000 and 2001 and partial deregulation and privatization of the telecommunications sector, Nigeria quickly made up for lost time, becoming the country with the fastest growing ICT sector in Africa. In many ways therefore, it presents an interesting scenario study of the development and diffusion of ICTs deliberately to promote socioeconomic growth.

Nigeria possesses many of the characteristics of a developing country. These characteristics, according to Weatherby, et al (2006), include: delayed modernization, poor communication facilities, high illiteracy level, lack of skilled personnel, large population, lack of capital and appropriate technology, and unequal distribution of wealth. While these features are common in many developing countries, for years they were compounded in Nigeria by other conditions such as religious and ethnic conflicts, militarism and an overbloated state operating alongside weak institutional structures. An interplay of these factors potentially posed major challenges to the application of ICTs as tools for socioeconomic development in the country. While the different African countries have unique developmental problems, an analysis of the ways in which Nigeria tackles its challenges and harnesses ICTs for development may provide policy directions for countries yet to fully engage with these technologies, or which are at the same level of ICT development as Nigeria.

Research Methodology

The book adopts a single-case study, multi-level, qualitative and quantitative approach. It spans different levels of analysis—society and state. While it is ostensibly a state-level work in the sense that it focuses more on the behavior of the state, particularly as it responds to the technological changes occurring at the systemic level, it is also a work that considers the societal framework. One proceeds from the premise, even if self-evident, that it will take more than the action of the state for ICTs to be successfully diffused and harnessed for socioeconomic development in any country. The success of the project will involve various state and non-state (societal) as well as other systems-level forces.

The case study approach, rooted in the work of Max Weber and German historiography, is historical, interpretive and holistic, and objects of study are taken as whole entities. It is concerned with the "intersection of a set of conditions in time and space that produces many of the qualitative changes" in a society (Ragin, 1982: 25). This approach is given the subject matter of this book for several reasons. First, it is widely used in information technology research because of its focus on the institutional framework, policies and intentions of policymakers. As Montealegre (1999) argued, "given the widespread prescription of IT for Third World countries, the urgency of their needs, and the paucity of their economic resources, it seems useful to understand the state of IT absorption in different countries and the effect of different technology levels, cultures, and priorities on IT implementation" (p.199).

Secondly, the approach enables the identification of factors that are crucial to the successful implementation of policies on ICTs. These factors include the state of the infrastructure, and the institutional and cultural frameworks. The research interrogates the connections between these forces (or variables) and ICT-related national policy goals. Thirdly, the approach, which is both descriptive and explanatory, enriches an understanding of the emergence of ICTs in Nigeria.

Finally, the fundamental question addressed in the book require a case-study method to effectively and qualitatively examine and interpret the extent to which Nigeria's policies on ICTs are expected to promote the stated goals of harnessing the technologies for socioeconomic development. It also enables a systematic and exhaustive examination of Nigeria's capacity to develop and harness ICTs for socioeconomic development given the infrastructural, institutional and cultural factors significant in the process of acquiring and diffusing the technologies.

The research involved several techniques. One of these was a questionnaire, administered to two groups of participants in 2001 and 2007. I also used personal interviews of key individuals in the ICT industry both in the public and private sectors. Content analysis of policy documents, participant-observation and anecdotal evidence about ICT usage and diffusion in the country were other methods of data collection.

The book is organized in eight chapters, including this introductory chapter. Chapter Two presents an overview of the relevant theories in the ICT4D field—from development theory to information society theory. While there are no specific theories of ICTs and development, as the chapter will show, many of the current assumptions about the connections between ICT acquisition and socioeconomic growth are grounded in various strands of development theory especially classical theories of modernization. Chapters Three and Four concentrate on Nigeria's history of economic development and its policy framework on ICTs. The contributions of the private sector are also discussed in Chapter Four.

Chapter Five begins the empirical section of the book with an investigation of the patterns of ICT usage in the public sector. It connects the policy framework to ICT practice by examining, through observation and personal interviews, the levels of ICT penetration and usage in selected federal government ministries and departments considered crucial to the development and diffusion of ICTs in Nigeria. The case study continues in Chapter Six with an overview of the patterns of ICT usage in the societal context. It focuses on the societal response to the efforts to diffuse and harness ICTs for socioeconomic development in the country. This response is assessed through a study of ICT practice, as well as perceptions and expectations about the technologies by a section of civil society at two different time periods, August to December, 2001 and January 2007. The research periods are significant because the first marked the conclusion of the deregulation of the telecommunications sector that began in 1992, and the next five years (culminating in December 2006) witnessed the phenomenal diffusion of ICT usage in the country. The research compares the change in practice, usage and expectations of selected Nigerians between the two periods.

Chapter Seven examines the factors that have mediated the harnessing of ICTs for socioeconomic development in Nigeria. I refer to these factors as "potholes" on Nigeria's access route to the information superhighway, and use "detours" for the measures that policymakers and other stakeholders in Nigeria's ICT industry have devised to get around them. Chapter Eight is the concluding chapter in which I attempt to position Nigeria within a continental context in regards to how its process and experience in ICT development can enhance its relevance in Africa.

Notes

[1] In his *Runaway World* (2000), Giddens seems to have a more benign view of the erosion of individuality in an increasingly organized society where planning depends on information about people. At least, that is the sense one gets from his uncritical definition of globalization as "simply the way we live." He refers to a 24-hour global marketplace, and a global age very different from past eras. Globalization, he argues, is a "total change." It "is much more intimately linked to our lives. It is a shift in relationships where the global intersects with the private (resulting in) increasing connections between local life and global change."

[2] One must note that Webster is very critical of the idea of the "information society" but used it for practical reasons as the title of his book, *Theories of the Information Society* (London, England: Routledge, 1995).

[3] Rajaee (2000) highlights the pivotal role of the WWW, created in 1989, in promoting the rapid effects on society that the Internet has. He notes that: "an important date for the information society and for civilization was 1989, with the invention of the World Wide Web, comparable in important to the industrial and agricultural revolutions for industrial and agrarian societies." (p.72)

[4] Nigeria was ruled by different military governments in two time periods: 1966-1979 and 1983-1999. A civilian administration, headed by a retired general, began in 1999 and successfully handed over to an elected president in 2004. That marked the first peaceful democratic alternation in the country.

• CHAPTER TWO •

Information and Communication Technologies and Theories of Development

Introduction

Development theory is defined as the body of theories that focus on issues of development in countries in the periphery. It offers "explanations for the different kinds of patterns of development and underdevelopment which have occurred in the Third World" (Martinussen, 1997: 8). To be considered a theory of development, a theory must provide concrete insights into a developing country's actual problems and prospects for change. It must also "account for the uneven pattern of development worldwide and to recommend measures to overcome underdevelopment" (Munck and O'Hearn, 1999: xiii).

Earlier theories of development were usually economistic, focusing on macroeconomic indicators of growth. However a purely economic analysis would not pass for a development theory unless it also offers explanations for "development, maldevelopment or underdevelopment and stagnation" in developing countries (Martinussen, 1997: 13). In other words, to be considered as a theory of development, the theory must apply only to the study of developing countries. Martinussen further suggests what may very well be the test of a development theory.

> A development theory seeks to answer questions such as the following: How can chosen and specified development objectives be promoted? What conditions will possibly obstruct, delay or detract progress towards the objectives? What causal relationships and laws of motion apply to the societal change process? What actors play dominant roles, and what interests do they have? How do the changes affect various social groups and various geographical regions? (p.14)

Theories of Development

Development theory emerged in the 1950s and 1960s, perhaps in response to the decolonization of the South. Proponents of development theory undertook to explain the differences between the economic and political situations of the "undeveloped" Third World/South and those of the "developed" nations of the North.

> The United States was at the height of its power and influence and regarded itself as both the inspiration and the policeman of the world. Anthropologists, sociologists, psychologists and economists all contributed to the modernization project with unquestioning optimism. Progress in its new clothes was not only inevitable, it was obligatory. ... Modernizing elites and comprador development theorists in 'developing' countries eagerly enrolled in the project (Tucker, 1999: 7).

According to Dordick and Wang (1993:16) "... massive postwar assistance by the United States, in particular, called for some guidance to ensure that the money would be well spent." Thus four major approaches emerged in the examination of the problems and conditions for development in the new states. These were:

> A theory that drew upon historical experience to formulate a linear process of growth (stage theory); one that placed structural reform at the top of the list for success; another that recognized that while old-style empire colonialism was dead, a new form of colonialism fostered by capitalism had taken its place and had to be dealt with to promote development among the less fortunate nations; and more recently, a theory that argues that a return to traditional market economics will solve the development problems of the developing nations. (Ibid.)

The first two theories fall under the rubric of modernization theories of economic growth, while the third is Marxist-based dependency analysis of underdevelopment. The last approach is what has often been referred to as neoliberal economic theory, especially as interpreted for developing countries by the Bretton Woods institutions, particularly the International Monetary Fund (IMF).

While contemporary theories of development do not clearly distinguish between economic and political development, modernization theorists attempted to differentiate between the two, stressing either the development of a capitalist economy or a democratic polity.[1] According to this approach, development was indicated by factors such as urbanization, political institutions, technological advancement and economic infrastructures. It therefore did not matter which goal a country pursued as the outcome was the same—economic and political development. Almond and Verba (1965), for instance, concentrated on the political culture of development as expressed through the establishment of political institutions, citizen mobilization and participation balanced by state order and authority. Their focus remained on the nation-state and on the assumptions that political beliefs are relevant to the problem of political development. While the expected outcome was the same, other scholars emphasised the development of economic policies, infrastructure and per capita incomes as measures of success. Indeed Walter Rostow (1961), the key proponent of the stage theory, outlined the stages of modernization through which each society or "developing nations" must pass. He presented modernization as an inevitable trajectory for all societies which must evolve from the traditional stage, through the preconditions for take-off and drive to maturity, to the age of high mass consumption. In essence, Rostow and other development theorists believed that the new states would advance from a stage of underdevelopment to one of capitalist democracy. In this conceptualization, development was understood as the "maximizing of goods and services available" and therefore a developing country could be distinguished from a developed one "by the paucity of these goods and services" (Dordick and Wang, 1993:22).

Implicit in most theories of modernization was the belief in the superiority of American or Western values, institutions, and processes. Modernization theory focussed on progress as leading inevitably to development as defined by the Western experience. For instance, it was argued that the instilling of certain Western political values was critical to, and intertwined with, economic development. According to early theorists, factors for underdevelopment included traditionalism, lack of achievement motivation, regime type and culture. For states to become modern and developed, they had to adopt Western ways and attitudes. Policy prescriptions were given accordingly and policymakers in many countries initiated projects aimed at transforming traditional values and attitudes into modern (Western) concepts. While this development agenda was highly criticized for its ethnocentrism (Mehmet, 1999), it nonetheless structured the development principles in decolonized countries. Different strategies, beginning with mercantilist orientations, import substitution industrialization, basic-needs strategies and structural adjustment programs, were adopted at various times, in response to Western-centered development imperatives. By the 1990s, the focus had shifted to ICTs as new tools for achieving old objectives.

Classical modernization theory relied on internal factors to explain underdevelopment. However, in doing so, modernization theorists disregarded the structure of the international political economy and the ways in which it adversely affected the development efforts of countries in the South. It was to correct this "oversight" that dependency theory emerged in the late 1960s and 1970s, and directed attention to external factors of underdevelopment. Dependency theorists such as André Gunder Frank (1967) and Dos Santos (1970) argued that underdevelopment is a consequence of the unequal relationship of exchange and dependence that exists between core countries and those in the periphery. According to them, dependency is a product of an active process of unequal economic power relationships between two countries or groups of countries, and it comes in three forms: colonial, financial and technological-industrial. Underdevelopment, from the perspective of dependency, can be summarized as follows: The very process that leads to economic growth and development in rich countries results in underdevelopment in poor, mostly formerly colonized countries through negative terms of trade, the debt trap and technological-industrial dependency.

Some assumptions of dependency theory re-surfaced in Castells's analysis of the factors that hinder development in South countries particularly sub-Saharan Africa (Castells, 1996: 83-95). According to him, the causes of the region's marginalization and poverty are attributable to a number of things. First, over-reliance on the export of primary commodities (92% of all exports), 76% of which are agricultural products; and depression of their prices and negative terms of trade which, "as a result of the structure of exports, make it extremely difficult for Africa to grow on the basis of outward orientation of its economies" (p.83). Secondly, adjustment policies which stress export-oriented strategy of growth have increased reliance on primary commodity exports, thus exacerbating the region's underdevelopment. Thirdly, the collapse of Africa's industrialization efforts coincided with the period when technological renewal and export-oriented industrialization characterized most of the world, including other developing countries. Finally, the disconnection of Africa from the global network society further marginalized the continent. Castells adds:

> Technological dependency and technological underdevelopment in a period of accelerated technology change in the rest of the world, make it literally impossible for Africa to compete internationally either in manufacturing or in advanced services…The disinformation of Africa at the dawn of the Information Age may be the most lasting wound inflicted on this continent by new patterns of dependency, aggravated by policies of predatory state (1996: 95).

While much has changed in the years since Castells made this observation, sub-Saharan Africa continues to lie in the periphery of the global network society. For instance, in 2007 Africa as a continent had the lowest number of internet users per 100 inhabitants, at 5.43, compared to 20.12 for the world (ITU, 2008). It also lags behind in other indicators of the global network society such as cellular phone usage—at 28.11 (49.83 for the world), though the continent records the fastest growth in cellular phone usage in the world.

Theories, policy prescriptions and initiatives based on a classical understanding of development have since moved in different directions. Rather than top-down economic policies, programs promoting decentralized patterns of development have become prominent, and emphasis has at various times shifted to projects which directly target the poor, especially in rural areas. One of these policies focused on basic-needs strategy as a more useful approach to achieving the kind of economic growth that benefits the majority of people in developing countries. It contrasts with a top-down development model in which "the process of economic growth is characteristically thought to follow a series of 'stages' which would ultimately spread benefits to all, thereby alleviating poverty and inequality" (Brohman, 1996: 201).

The basic-needs approach targeted equitable distribution of resources, poverty alleviation and the satisfaction of basic needs through the adoption of appropriate technologies, rather than rapid modernization (Brohman, 1996, Dordick and Wang, 1993). It particularly aimed at de-linking development from economic growth as there were indications that macro-economic growth did not always trickle down to the majority of the populations. It became part of the international development agenda in 1976 when the ILO adopted the Declaration of Principles and Program of Action for a Basic Needs Strategy of Development. The ILO defined basic needs to include: minimum requirements of private consumption (e.g. food, shelter, clothing); essential services of collective consumption (e.g. electricity, water, sanitation, health care, education, public transport); and participation of people in decisions affecting their lives (Brohman, 1996: 205). The ILO focused on "harnessing local resources and providing the poor with the means to fulfil their development potential...It also acknowledged the need for structural (internal) change in the development patterns of Third World societies to meet the basic needs of the poor" (Ibid). While the basic-needs approach to development was well received by the community of development scholars and practitioners, before it could bear any fruits, it was quickly replaced by the structural adjustment programs that came on the heels of the debt crisis that plagued many developing countries in the 1980s.

Prior to the structural adjustment program taking the center stage of development policies in countries in the periphery, Africa had actually started to implement a basic-needs strategy of development that put the needs of its peoples atop the agenda of development. In 1980, 50 African heads of state convened in Lagos, Nigeria under the auspices of the Organization for African Unity (OAU) to discuss the way out of the continental affliction called poverty. At the end of the conference, the heads of state signed the Lagos Plan of Action (LPA), which essentially aimed at de-linking from the world economy to focus on internally driven strategies for development and collective self reliance. The plan called on the continent to:

> use its extensive resource base primarily for its own development rather than for export, to expand its industry primarily for home consumption and only secondarily for export, to rely principally on its own technical skills and not on those of foreigners, and to develop an industrial base and consumption patterns suitable to African needs and customs rather than blindly adopt models from abroad (Browne, 1984: 803).

The LPA was perhaps motivated by an overt understanding of dependency's assumptions, namely that reliance on export of primary products impoverishes peripheral countries. Within the first two years of the Lagos summit, steps were taken toward a self-sustaining African economy in which activities would be geared toward meeting the needs of Africans. These initial efforts included the mandate for the African Development Bank (ADB) to provide funding for cooperative and regional endeavours in line with the LPA. Existing regional organizations such as the Economic Community of West African States (ECOWAS) were strengthened and new ones created to foster intra-African and inter-regional trade. Preferential trade areas were established and trade barriers were eliminated to ease trade between African countries. Browne (1984) notes that though the LPA aimed at reducing Africa's dependence on other countries—"delinkage if you will—it certainly (did not) call for autarky. It is not a demand for Africa's total economic isolation from the rest of the world. Rather, it is a call for the Africans to shift away from reliance on them and a greater reliance on themselves" (Ibid).

The call came shortly before the African debt crisis and the consequent involvement of the Bretton Woods institutions in African economies. The IMF conditionalities were antithetical to the objectives of the LPA with their focus on external-oriented development strategies such as intensive production of primary products for export. The thrust of the structural adjustment programs in African countries was on the generation of sufficient foreign earnings to satisfy foreign loans. These countries therefore not only sank further into the debt trap, but continued their reliance on exports of primary products at prices set by Western buyers. In many cases, several countries were exporting the same agricultural commodities to an already saturated overseas market. Earnings from these activities went into debt servicing and whatever was left over was applied to the purchase of capital and consumer goods at prices equally set by industrialized countries. The vicious circle of dependence thus continued, validating many of the assumptions of dependency theory.

Dependency theory being intrinsically in opposition to the Establishment—neo-liberal international financial institutions—soon faded and modernization theory resurfaced in the 1980s and early 1990s in response to the perceived "miracle" of the Asian Tigers (which at the time indicated that South countries could succeed). The period also spawned scholars of diverse theoretical leanings who formulated such notions as complex (or mutual) interdependence which implied that a dependent relationship can be mutually beneficial to all the actors (for instance, Keohane and Nye, 1977). Tracing what has become known as reverse dependency, these scholars argued that industrialized nations are dependent on the developing countries for raw materials, and for markets for their exports. But as some other scholars (such as Brohman, 1996; Martinussen, 1997) have shown, the industrialized countries are not impotently dependent on poor countries. For one, while developed countries have many choices of countries from where to import and at prices they set, poor countries are dependent on a limited overseas market for their low-value primary goods. Secondly, with their vast array of technologies, industrialized countries are constantly developing synthetic and alternative raw materials to replace imports from the South. One example is the decreasing reliance on textiles because of the production of synthetic fabrics such as polyester to replace cotton and other natural fibres in clothes manufacturing.

Modernization Theory of Development Communication

While the primary focus of modernization theory was on economic and political development, there were scholars who studied social relations through the linkages between information/communication and development. They argued that traditionalism fostered resistance to progress and modern attitudes, and was therefore a major obstacle to the goal of maximizing goods and services. The solution was a change in values and attitudes. The

> ... quickest and most effective way of bringing about this change of consciousness was the application of "technology-based" communications, principally radio, and in time, television. Literacy is not required; images are created that promote "psychic mobility"; and messages concerning health, education, farming methods, and more are delivered (Dordick and Wang, 1993: 22).

According to Lerner's exposure theory of communication, contact with modern values through the mass media transforms behaviour and attitudes and in the process creates a political and economic actor (1958). This actor was expected to be the agent of economic growth because he or she would sow the "right seed, use credit efficiently, voice political views and demands through the appropriate channels, and organize the institutions needed to push traditional societies over the threshold of modernity and into the twentieth century" (cited in Mowlana, 1997:188). The actor's modern attitudes and influence would, again through the media, diffuse to the rest of the traditional society and its people would adopt modern attitudes about savings that would in turn usher in economic development. Classical and neoclassical theorists therefore perceived "communication as a necessary factor for economic development and growth" (p.189). As Lerner further argued, there was causality between communication on one hand, and urbanization and education, on the other. Together, they resulted in development. Pye (1963, 1965) added that communication was a prerequisite for development because of its power to destroy traditional societies. Inkeles and Smith in their famous work, *Becoming Modern* (1963), enthused about the role of the mass media in effecting modernization. The media, they argued, were the inculcators of individual modernization.

This was the beginning of development communication, defined as "the intersection of the communications and the economic and social sciences" Dordick and Wang (1993) and its deployment as a strategy of economic growth for developing countries. Dordick and Wang (1993) speculate that development communication might have arisen as a result of the failure of the purely economistic stage theories of the 1950s. Scholars therefore turned to

> the miracle of electronic communications as a means for stimulating growth in the lesser developed countries. And indeed communications technology appeared to have promise. The transistor emerged from the laboratory and rapidly appeared in the far-flung corners of the world as the transistor radio ... Television was rapidly becoming a household appliance in the developed nations, and almost every developing country was debating not whether but how they should introduce television with its power to inform, persuade, and educate, and thereby facilitate development. (p.20)

These arguments were premised on the assumption that there was a correlation between underdevelopment and the lack of information, especially as disseminated through technologies of the mass media. This meant that one could tell the difference between a developed country and an underdeveloped one by simply looking at the number of western-type ICT gadgets that each country had, as this was an indicator of how well informed the people were. This was the basis of Stover's assertion that "poor countries have fewer means of communication than rich ones, and the lack of information correlates with a low level of development" (1984: 8). He did not critically analyze his data to show the causality between the level of development and the presence of mass media gadgets. For instance, did the countries have high levels of "means of communication" because they had high purchasing power and could afford them or, did the acquisition of these "means of communication" enrich them? As the discussion on the productivity paradox[2] indicates, even in industrialized countries, there is no evidence that investment in ICTs directly results in increased economic productivity. Even though "more recent studies indicate positive productivity findings" (Avgerou (1998) and the presence of ICTs has expanded employment opportunities for people in developing countries, the evidence is still scanty on the causality between investments in IT and economic growth.

Modernization Theory and the ICT-for-Development Discourse

Classical modernization theory has since evolved to adapt to the new challenges confronting developing countries (So, 1990). However, its original assumptions overtly or implicitly structure much of the current debates linking ICTs with development. This is despite the fact that the theory of development communication, like many classical modernization theories, has since proved inadequate in explaining the problem of underdevelopment in sub-Saharan Africa (So, 1990; Brohman, 1996; Martinussen, 1997 and Seligson, 1998). As Dordick and Wang (1993: 25) note, the idea of development communication "is the forerunner of current thinking about *development informatization*." (Italics are in the original).

Even as these authors acknowledge the historical linkage between development communication and development informatization, they however ignore the point that even in the 1950s and 1960s, *information* was integrated in the assumptions made by theorists of development communication. Back then, there was no conceptual separation between information and communication and their role in the development process. The focus was not merely on the means of communication (the technologies) but on the messages that were communicated (information). Development informatization—a process of change toward an information society—invokes notions of the prevalence of information as a factor of production (and therefore economic growth) in ways that are different from information starkly conceived in the early days of development communication as having the capacity to transform attitudes. However, regardless of the differences between the roles of information in development then and now, the classical assumptions about information today inform current policies and practice in the acquisition and use of ICTs in developing countries.

ICTs and Development: Emerging Theoretical Perspectives

There are no explicit theories of ICT in the context of development, though the field is not in short of superlatives on the wonders of the new technologies. The literature is replete with prescriptive accounts of either assumptions about the potential of ICTs in development or case studies of micro applications of ICTs in developing countries—such as Bada's work on ICT usage in one Nigerian bank (2002). As Heeks (2006: 1) points out,

> There has been a bias to action, not a bias to knowledge. We are changing the world without interpreting or understanding it. Most of the ICT4D research being produced is therefore descriptive not analytical. It might make some interesting points but it lacks sufficient rigor to make its findings credible and it can often be repetitive of earlier work. It has close-to-zero shelf life.

Heeks argues that theorizing in ICTs and development (which he refers to as development informatics because it is "less technocentric") has drawn from work in library and information sciences, communication studies and information systems without much input from development studies, a key prong of the ICT4D discourse. Even the approach that he proposes still draws from other disciplines ("we set the whole of social sciences as our boundary," [2006: 3]) in a four-part format aimed at providing rigor and credibility to research in the field. The focus on the D-part of ICT4D (development) was also the choice of participants at an IDRC workshop in 1997 when they developed what approximated a systematic theory of ICT and development articulated from an analysis of the "five indicators of development" (Howkins and Valantin, 1997).

These indicators are: *literacy*, education, and skills (literacy, education, training and skills, and opportunities for all members of society to increase their capacities); *health* (life expectancy, maternal and infant mortality, quality of life, and the levels of health care available in situations of morbidity); *income and economic welfare* (high levels of employment, high incomes per capita, and increased gross national product, with appropriate corrections for environmental protection and for income equity); *choice, democracy, and participation* (participation in social and economic affairs, with fair economic rewards, the availability of reasonable choice, and participation in the democratic process); and *technology* (the capacity to develop technological innovations and to make technological choices (Howkins and Valantin, 1997). Most assertions about the capacities of ICTs to promote development are made in the context of how they improve on these indicators. For instance, Mungai (2002) presents case studies of the ways the internet is used in different African countries in the areas of medical consultations, data transmission, appointment scheduling, telemedicine and information on specific diseases such as AIDS/HIV and meningitis. Hamelink (1997) also argued that ICTs can be used for education, health, environment, agriculture and small-scale industrialization in ways that will contribute to general economic growth. He summed up the different positions held by advocates of ICTs for development in a 1997 UNRISD discussion paper, identifying two major categories: techno-centric utopian and dystopian/pessimistic perspectives.

Dordick and Wang (1993) use the concepts of "band wagon approach" and "fortress approach" to refer to the same distinctions. Heeks (1999) explains the same phenomenon but through a different taxonomy in which he identifies two continua: technology impacts (from optimists to pessimists) and impact causes (from technological determinism to social determinism). Generally, utopians (or "bandwagoners") are optimistic about the abilities of ICTs to create wealth, employment and increase productivity, and raise economic growth in developing countries. They believe that ICTs represent a revolutionary and historically unprecedented force that can fundamentally transform societies and individual lives, and thus policymakers and elites in developing countries should hastily acquire and integrate ICTs in their strategies for development.

> The techno-centric perspective holds that the "digital revolution" definitively marks the passage of world history into a post-industrial stage. The emerging global information society is characterized by positive features: there will be more effective health care, better education, more information and diversity of culture. New digital technologies create more choice for people in education, shopping, entertainment, news media and travel (Hamelink, 1997).

This perspective echoes the concepts of revolution and discontinuity that pervade the discourse on ICT and development. In the literature, both in the contexts of developing and developed countries, there is a sense of urgency to leverage the transformative capacities of the technologies. In practice, policymakers, such as those encountered in Nigeria in the course of researching this book, also conceptualize ICTs in the language and posture of revolution—of the violent kind. For instance, a major stakeholder in the country's ICT sector, during an interview, repeatedly used words such as revolutions and weapons:

> This is a revolution. It takes time for you to see the fruits of the revolution. ICT is seen as a tool of the revolution, the jetfighter, the armoured car, the bombers of this revolution. Can you see? The ICT is a tool, one of weapons. It is not the end of the war or the revolution. At times our brothers from the West tend to mix the two. No,

we see ICT as a tool as one of the machine guns to be used, as the landmines or whatever is the most powerful weapon to use.

Hamelink (1997) argues that technological processes are rarely revolutionary. As noted in the above quote (even if it contradicts the general idea of revolution), their effects are more gradual than they are radical breakthroughs. The dystopian/continuity approach to ICTs is mostly drawn from extant research on the social impacts of ICTs in developed countries. This more critical analysis rejects "the idea of discontinuity and stresses the likelihood that ICT deployment will simply reinforce historical trends toward socioeconomic disparities, inequality in political power and gaps between knowledge élites and the knowledge-disenfranchised" (Hamelink, 1997: 27). The complexity in reaching theoretical unanimity on how to explain and understand the process of ICT for development is resonant with a similar challenge in the study of the information/network society. Indeed, the jury is still out on whether there is a global information/network society at all despite the unequivocal proclamations of Castells (1985, 1994, 1996 and 2003) and Friedman (2000, 2005) about its emergence. The discourse over the existence and nature of the information society is important in the context of developing countries because the literature on ICT4D stresses the significance of ICTs as tools to achieve the twin objectives of generating internal socioeconomic growth in developing countries, and accelerating their integration into the global network/information society. It is therefore necessary to sketch some understanding of the information/network society towards which African countries are journeying.

Defining the Information Society

The notion of the "information society" is rooted in ideologies of improvement manifested through two concepts. The first derives its theoretical origins from the "writings of Rolf Dahrendorf, Daniel Bell, Jacques Ellul, and others, who sought to relate increasing sophistication of technology and planning to the emergence of a new society" (Dordick and Wang, 1993: 9). Three "paradigms of progress"—the new liberty, post-industrial society and technological society—dominated the works of these writers. Dahrendorf (1975) particularly focused on the arrival of a new society in which liberty would flourish following the satisfaction of material needs. Bell (1973) conceptualized the post-industrial society as the logical stage after Walt Rostow's age of high mass consumption. In this new society, "the acquisition and codification of theoretical knowledge" become the driving economic forces in the society, replacing the manufacturing sector. This society, which subsequent writers have called the information society, is characterized by the growth of information and knowledge-based industries. Bell's thesis was that "there was a change from the production of goods to the provision of services, and that this underpinned the production of knowledge" (Dearnley and Feather, 2001: 14). While Dahrendorf wrote about the aspiration to political goals and liberty as the next stage (following the satisfaction of economic needs) and Bell envisioned a post-industrial society in which information and knowledge would become the driving factors of production, Ellul (1964) wrote about a technological society where there would exist the unification of man (sic) and technique.

The second concept in the formulation of the information society emerged through the works of Tadeo Umesao and Masuda (in Japan); Machlups, and Porat and Rubin (in the United States). These writers focused on the ubiquity of information and knowledge and compared their economic significance to the production of goods that prevailed in the capitalist mode of production and has existed since the Industrial Revolution. The concept first featured in the writings of Japanese intellectuals through studies begun in the 1960s at the Japan Computer Usage Development Institute (JACUDI). Masuda compiled the result of these studies in a compendium (Masuda, 1981). He concluded that the "imperatives of progress and the maintenance of human values" could be harmonized but in ways that would allow a replacement of material values by more spiritual ones.

> He proposes a society in which information values, rather than material values, are the driving force. He points to economic factors that constitute this society: universally available information at affordable costs, and quantity and quality of information with facilities for the distribution of the stocks and flows of this information. As a result information communities on a human scale, participatory democracy, and the spirit of globalism would emerge (Dordick and Wang, 1993: 13).

Fritz Machlup (1962) was one of the earliest to observe the shifting work force from the agricultural and manufacturing sectors to information and knowledge production. And he argued that this production of information and knowledge was as economically significant as production of goods. He did a major study of the US economy in which he

> examined expenditure on such matters as education, research and development, broadcasting and legal services...(and argued) that all these forms of production were knowledge-related; they were either dependent upon knowledge (education), or were a form of knowledge dissemination (publishing) or, crucially, were concerned with production of new knowledge (research and development) (Dearnley and Feather, 2001: 12).

Machlup's work was pivotal in prompting the American government to commission (and publish) an analysis of the information economy by Porat (1977). This contributed further to an understanding of the information society.

Given such origins, it is not surprising that most definitions of the information society are economic and technological. They emphasize the role of ICTs in creating a knowledge-based economy; the preponderance of knowledge-based industries becomes an index of progress. Definitions from these perspectives also highlight the contribution of the knowledge/information sector to a country's gross national product—in comparison with other sectors of the economy. But the information society has also been defined in non-economic terms. For instance, Webster (1995) suggests that the information society is commonly defined from a technological perspective with emphasis on how "breakthroughs in information process, storage and transmission have led to the application of information technologies (IT) in virtually all corners of the society" (Webster, 1995:7). This perspective stresses the falling costs, miniaturization, increasing power and pervasiveness of ICTs in the society such that as Ellul might suggest, there is no longer a distinction between "man and technology." A third approach to the definition of the information society is occupational as seen in the work of Porat (referred to earlier). It concludes that "we have achieved an 'information society' when the predominance of occupations is found in information work" (Webster, 1995: 13).

The information society is further defined spatially to stress the compression of space (and time) engendered by the information networks of ICTs. These networks connect people in distant locations to create a global network society (Castells, 1996). The network society refers to movements, linkages, and flows which reshape and often undermine the integrity and coherence of borders and spatial entities. Network, a word spun from the actions of a spider and its web creation, refers to any system of crisscrossing lines, or a group of people who collaborate informally to promote common goals. Technologically, a network connects several computers together. In the ICT-enabled network, parts become intrinsic pieces of the whole to a greater degree than ever before.

According to Castells, networks now constitute the new social structure of our societies as the diffusion of networking logic modifies the processes of production, experience, power and culture. Information flow is emphasized in this definition of the information society, with the information networks usually compared to the electricity grid in a way that implies total connectedness. The "rise of the network society" (the title of one of Castells's books) has led to an increase in the trans-border flow of information, globalization of finance, or the cross-border flow of credit (such as loans and bonds), investment, money (foreign exchange) and other financial instruments, in unprecedented degrees. The information networks facilitate the internationalization of production such that they are considered major drivers of globalization (Friedman 2001). Outside the economic realm, the information networks are also connecting people of diverse origins, thereby creating an arena through which a global civil society is emerging.

Finally, a cultural definition of the information society logically flows from the above. With the increase in information through the new channels of communication, people are more aware of each other and the different cultures. Television remains the prime agent in cultural diffusion. And these days, with satellite and cable communication systems, distant places are beamed to living rooms around the world—the part of the world where the availability of electricity and television sets is assured. Movies, rental videos, printed materials (such as novels and magazines), E-books and electronic libraries and databases convey cultural messages from one geographical location to another. Culture has gone transnational with porous borders and movement through time and space. Messages and meanings have been recorded, preserved and reproduced. Individuals have also become important carriers of cultural practices through advances in transportation technology.

Obviously, there are overlaps in the technological, economic and occupational definitions of the information society. Also, since many of these definitions obviously proceed from a utopian perspective of ICTs, there is no interrogation of the disparities between peoples and countries in the process of informatization. For instance, in using information flow as a determinant of information society, the issue of unidirectional flow of information is often overlooked, even by more critical analysts. This is an aspect that developing countries often overlook as they formulate and implement polices aimed at generating internal economic growth and accelerating their journey into the information society. Despite these problems however, theories have emerged to explain the information society (in its many definitions) and its consequences for the industrialized countries that are already part of the process, and the developing countries on the margins.

Theories of the Information Society

In the following section, I survey some of these theories and the various approaches to an analysis and understanding of the information society and ICTs in the context of the prospects and challenges that the technologies present to a country like Nigeria. This exercise is expected to illustrate how some of these theories may be used to study the implications of ICTs for development in a manner that is more critical, and thus constructive, than the prevalent utopian assumptions that pervade the popular discourse on ICTs and development. For sure, one can find some critical analyses of ICTs even in the context of developing countries. Morales-Gomez and Meleese (1998) acknowledge the emergence of a "growing voice, albeit very small, in the current discussion on ICTs." There are many approaches to the study of the implications of the information society, but only four of these are relevant to the current purpose: stage approach, scenario-modelling, structuralism and critical theory (or political economy approach).

The Stage Approach

Analysts who adopt the stage approach seek to explain the levels of penetration and diffusion of ICTs in a given society. Kendall (1999) adopts this approach in the analysis of ICTs and their implications for development. He identifies five stages in the life cycle of technology: technological invention or discovery, technological emergence, technological acceptance, technological sublime and technological surplus. Many African countries, including Nigeria, are clearly at the third stage of ICT life cycle in some technologies (such as cellular phones) and lower in others. While a stage approach may help in articulating the levels of ICT diffusion in developing countries (and will be incorporated in the analysis of ICT diffusion in Nigeria in Chapters Five and Six), it is technologically deterministic. Also, it does not account sufficiently for the processes through which ICTs can lead to socioeconomic development in these countries. It posits technology as a necessary condition for economic growth with the implications that countries just need to progress from one stage to the next. Also, like all stage theories, this approach is evolutionary and therefore creates certain difficulties as Webster (1995) notes. "Evolutionist thinking has been out of favour for a long time in social science circles" because of its intellectual vulnerability to "at least two serious charges...the fallacy of historicism (the idea that it is possible to identify the underlying laws or trends of history and thereby to foresee the future)... (and the) trap of teleological thinking (the notion that societies change towards some ultimate goal)" [p. 35].

The stage approach also fails to address the inherent "underside" of technology, or the contradictions that technological diffusion necessitates. It implies that all countries can move from one stage to the other merely by acquiring and using the technology without considering the socioeconomic implications or the factors that might intervene in the achievement of stated goals. Already, certain unintended consequences of "technology" (as understood in the technological transfer debates of the 1970s) have proved counterproductive to overall national economic goals in some of these countries, particularly those in sub-Saharan Africa. On the other hand, the interaction of these technologies within the varying contexts of African societies has created new forms and functions of ICTs distinct from their conceptions at the site of production (Akpan-Obong, 2009).

The earlier argument was that technology could be transferred seamlessly from the North to the South—from the site of production to the context of usage. But as Makinde (1986) argued, the whole notion of technological transfer was flawed *ab initio*. In the first place, "technological transfer (implied) its total disappearance in one place and reappearance in another without any modification" (p. 177). The concept also connoted business, "thus technology transfer is merely another term for selling and buying technological products." Finally, technology transfer implied a "circulation of technology" with a "reciprocal flow of resources from one nation to another, like the dissemination of knowledge" (Ibid). This did not occur because it was not the technology—as in the knowledge and skills—that was transferred, but the finished products. The result was that many developing countries transformed into consumers of technology—the finished products—which in many cases was not appropriate to their local needs. In other cases, the technology was obsolete, overpriced and the notion of transfer actually implied the relocation (or dumping) of environmental liabilities in developing countries (Castells, 1996; Martinussen, 1997).

Scenario-Modeling

Another approach to the analysis of the information society (as well ICTs in the context of developing countries) is scenario-modeling or foresight study. Analysts (and policymakers in some cases) conceptualize possible outcomes based on certain conditions or scenarios. They create a scenario in which the presence of specific variables is expected to lead to certain predetermined results. South Africa commissioned such a foresight study in 1997, in which the government set up various initiatives. "After a wide-ranging process of selection, working groups of 20-30 people were formed to represent diverse interests and experience in each of 12 sectors" including ICT (Miller and Day, 2000). The objective was to predict the possible outcomes if certain initiatives were implemented in the ICT sector. This approach is also popular among futurists and environmentalists. Hammond (1998) used it to predict the outcome of the current process of globalization.

Of more relevance to the issues raised in this book is the way in which Howkins and Valantin (1997) apply scenario-modeling to an analysis of the implications of ICTs in a manner that integrates both elements of development theory and information society theories. Thus they provide a conceptual linkage between the effects of ICTs on industrial countries and on developing countries. They model four scenarios in the use of ICTs as development strategies: the March of Follies, the Cargo Cult, Netblocs and Networld.

Scenario 1–March of Follies: The global community is exclusive and fragmented and most developing countries tend to "respond only partially and reactively to the use and acquisition of ICTs" (Howkins and Valantin, 1997: 29). In this scenario, the market is competitive and cooperation manifests only in mergers and concentration to reduce transaction costs and maximize profits.

Scenario 2–Cargo Cult: Most developing countries assume that the global community is inclusive and supportive, but they respond only partially and reactively to the acquisition and use of ICTs. Howkins and Valantin argue that the cargo-cult mentality of the 19th century prevailed in the debates of the 1990s about the potential of ICTs to transform developing countries. The new "religion," symbolized by computers, makes uncritical assumptions about what ICTs can do. Many developing countries, sometimes aided by international development agencies, are struggling to place computers in schools "even if they do not work well or have any useful software" (p.37). Some of these countries have become dumping grounds for antiquated ICT-related equipment phased out in Western countries, raising concerns from environmentalists about the future effects of this development on the environment. Many countries seem to be replicating "the import substitution strategies that had been popular in the 1950s, 1960s and 1970s" (Ibid). And just as those strategies did not result in development for many, the rush to wire developing countries will be abandoned, according to Howkins and Valantin, when ICT policies "fail to deliver goods and services that could compete with foreign products" (p. 37).

Scenario 3–Netblocs: Moving from the Cargo Cult, the world slides into regional blocs. In this scenario, the major assumptions are that the global system is exclusive and fragmented, and therefore "developing countries take an active approach to the acquisition and use of ICTs and develop a complete set of policies" (p. 38). But these policies lead to a world of regional blocs even as many people and countries are wired to the global information society. The emergent blocs or groups are based on shared cultures and languages with each bloc pursuing competitive economic goals without much cooperation with countries outside the blocs. Netblocs does not assume that all countries will belong as the lack of resources will exclude some. Others will be isolated for lack of "natural partners." This scenario eventually collapses as differing "regional laws, regulations, and trading principles create centripetal forces that lead to a highly unstable situation" (Ibid).

Scenario 4–Networld: This is the ideal destination of the journey from agricultural societies through industrial to post-industrial. In this scenario, the global community is perceived by all to be inclusive and supportive. "Developing countries have a complete and proactive set of policies toward the acquisition and use of ICTs" (Howkins and Valantin, 1997: 41). They treat information and communication as the starting point for development. Multilateral corporations—particularly those in the ICT sector—act in enlightened self-interest and pursue collaborative efforts with companies and institutions in developing countries. They persuade their governments to "dismantle trade barriers" and developing countries respond to this gesture with a realization that they should "work with global corporations to create their own national information society and economy" (Ibid.). Howkins and Valantin do not predict that a Networld will eradicate poverty and deprivation, but they argue that it will create a more supportive and knowledgeable environment than the other scenarios. While they do not ostensibly privilege any of the scenarios as being the most or best possible outcome, there is an implicit preference for the Networld. This preference undergirds many of the arguments about the information society—both in the context of industrial countries and the potential that ICTs offer for socioeconomic growth in developing countries. Like similar arguments—as Howkins and Valantin acknowledge—there is no direct linkage between ICTs and development to demonstrate how countries can move from the first three scenarios to the fourth.

However, the Cargo Cult may be closer to the reality in many sub-Saharan African countries as they encounter the new crusade of ICT. This scenario explains the policies and acquisition of ICTs as development tools in many developing countries, including Nigeria. Policymakers and other stakeholders in the county share the view that the "global community is inclusive and supportive" (Howkins and Valantin, 1997: 33). Particularly, the national policy on information technology, according to its vision statement, was aimed at making "Nigeria an IT capable country in Africa and a key player in the information society" with IT "as the engine for sustainable development and global competitiveness" (Federal Republic of Nigeria, 2001). This and similar policies were expected to guide the development and diffusion of ICTs as tools for economic development in the country. While the policy on information technology has provisions for indigenous development of the technologies and emphasizes research and development, the implementation has often leaned toward importation as a fast route to technology diffusion.

Toward a Theoretical Framework of ICTs and Development

While each of the different theories—either in the context of developing or industrialized countries—contributes to an understanding of certain aspects of the ICT4D discourse, none, by itself adequately explains the multi-faceted nature of the subject. An integrated framework that facilitates the explanation and analysis of the research findings becomes necessary. This framework "explicitly derives...from a body of (different) theoretical...perspectives" (Heeks, 2006: 3) such as those reviewed in this chapter.

Constructing an Alternative Framework

Kendall's stage approach is useful but only in the understanding of the status of ICTs—or penetration level—in Nigeria. The five stages in the life cycle of technology will be used alongside a framework of analysis developed by Peter Wolcott (1997) in quantifiably assessing the status of ICTs, as well as the level of usage in Nigeria. In the discussion of Kendall's stage approach, I observed that many sub-Saharan African countries, especially Nigeria, are in the third cycle, namely technological acceptance. As will be shown in the empirical chapters, there is an awareness, acceptance, and application of the technologies as tools to stimulate economic activities in Nigeria. But the spread is uneven, concentrating mostly in major cities. Even in these cities (especially Lagos and Port Harcourt) with relatively high percentages of urban poor, not many people have access to the technologies (besides the cellular phone). Thus a greater percentage of the population is likely to be excluded from the perceived benefits of these technologies.

Howkins and Valantin's scenario-modelling is useful in making tentative predictions of the process and likely outcomes of the development of the ICT sector in Nigeria. The approach also highlights the actions/activities of the state, as it is premised on the role of the state (and perhaps corporations) as active agents in determining the success or failure of the ICT4D project.

Conclusion

This chapter has demonstrated the ways in which classical theories of development, particularly those concerned with development and communication or information, have structured current debates on the connections between ICT and socioeconomic growth in developing countries. This has succeeded mostly because there is yet no articulated theory of ICT4D even as various scholars and researchers have drawn from various disciplines in their work in the field. Similarly, I apply an integrated framework of various perspectives in my examination of the process of ICT development and their utilization as tools for socioeconomic growth in Nigeria. Some of these perspectives will facilitate the analysis of different aspects of the research while others will be useful in explaining others. It is hoped that by focusing on the role of the various sectors and the institutional and ideological structures in the Nigerian society, the approach will provide a better explanation of the phenomena under discussion.

Notes

[1] In neo-liberal economic logic, there is seemingly no separation between economic and political reforms and this probably explains why the IMF, in handing out its structural adjustment packages to countries in economic crisis, stressed democratization as a precondition for economic development.

[2] The productivity paradox refers to the debates in the 1990s on the connections between presence of information technology and development, or prevalence of information workers and growth. In their research, Dordick and Wang (1993) found that countries with similar sizes of information sectors did not have the same level of growth. "For example, in 1987 Trinidad and West Germany both had information sectors between 32% and 33% of the work force, but Trinidad's GNP per capita was about $5,000, while West Germany's was about $14,000." (p.49-50) A correlation is yet to be established between high levels of ICTs and economic growth in developing countries.

• CHAPTER THREE •

Evolution of Nigerian Economic Development

Introduction

Once upon a time, Nigeria's economy was so buoyant that a head of state reportedly declared that the country's problem was not with money but how to spend it. And legend has it that in those days, Nigeria once went to the aid of a small West African country with funds to pay several months of arrears of salaries to its civil servants. It was also the days of huge infrastructural development in Nigeria, with many of the major roads and airports in the country built at the time. Nigeria was poised for the big league and everyone took it for granted that the "giant in the sun" would continue on this path of unimaginable opulence. This confidence was not unfounded—at least so it seemed at the time.

First, it was the 1970s—the decade of the oil boom (or oil shocks for countries that had to pay for the spiralling price of petroleum products when oil producing countries reacted to the Arab-Israeli conflicts by cutting oil production). Though oil had been discovered in Nigeria in the late 1950s, it was only around this period that the product was found in huge commercial quantity. And soon, it had become Nigeria's major foreign exchange earner—displacing cash crops, which for decades had been the mainstay of the country's economy. With seemingly inexhaustible oil wells and good price for the product, Nigerians—especially their leaders—were certain that the wealth had come to stay. Second, besides the new oil-generated capital, Nigerians had reasons to be confident because from all accounts, the country was abundantly endowed with two crucial factors of production: land and labor.

Nigeria occupies a relatively small landmass—923,800 square kilometres—but approximately 80-90 % of it is arable land, and rich for various forms of agricultural activity (Yesufu, 1996). Agricultural production is diverse, varying according to the pattern of climatic conditions in different regions of the country. The southern parts, with ample rainfall, grow food trees such as plantain and banana, and root crops such as cassava and yam, while the drier north region produces grains and livestock. The country is also rich in human resources: 140 million Nigerians, according to the 2006 national census. This population is composed of about 250 ethnic and linguistic groups with the major ones being the Hausa-Fulani (in the north), the Yorubas (in the west), the Igbos and the Annang/Efik/Ibibio (in the east and southeast respectively), the Ijaws in the Niger Delta region and the Kanuris in the northeast. The density of this population is unevenly distributed around the country with many of the northern states being sparsely populated while some of the southern states such as Lagos are densely populated. Lagos alone has an estimated population of 10 million, and is ranked among the most populous cities in the world.[1] The unevenness in Nigeria's population density is often traced to certain historical, physical, climatic, social and economic factors that are widely believed to be "mainly responsible for the present great disparities in the levels of development between geographical regions of Nigeria, particularly between the North and South" (Olaloku, 1979: 3). For instance, while Northerners also specialize in agricultural production, the frequent drought circumscribes how much food can be cultivated for sustenance. A large portion of the population (namely the Fulanis) therefore lives a nomadic lifestyle.

Nigeria is thus abundant in land, labor and the potentials for entrepreneurship and capital (with its natural and human resources). The question then is: why has the country slid from the level of middle-income country in 1979, by World Bank standards (Zartman, 1983:14), to its current status as a low income country, defined as one with a GNI per capita of $935 or less (World Bank, 2008)? This book is not designed to pursue the question of Nigeria's underdevelopment *per se*. However, establishing the potential role of ICTs requires some understanding of where the country has been.

The Nigerian Development Dilemma

Quite a bit of research has been done aimed at explaining Nigeria's underdevelopment. Yesufu (1996) provides a succinct summary of the various problems identified with besetting efforts at development in Nigeria. He makes a distinction between development and economic growth and argues that Nigerians have been "suffering from the illusion of economic growth without development." He considers the different stages of development in the country and concludes that while the macro-economic indicators show evidence of some form of economic growth, there has been no development—which he defines simply as "the increase in the welfare and quality of life of the citizen."

The lack of development in Nigeria, according to Yesufu, arises from several factors.

First, on the surface, the creation of states from three regions in 1960 to the current 36 and one Federal Capital Territory could have been a positive measure to launch centres of development and governance at the grassroots. Instead, it has resulted in "higher tendency of insularity, inward-looking and self-centred economic interests. Interstate frictions and mutual hostility have multiplied, and intrastate community and tribal squabbles have tended to get out of hand (and economic) barriers have hardened" (Yesufu, 1996: p.40). Uwadibie (2002) however faults the argument against state creation noting that the action led to the decentralization of governance with the purpose of devolving power and control to the grassroots and thus promoting economic development. He considers this to be a positive development because "devolution of authority from the federal to state and local governments" brought people-centred outcomes such as "a more effective implementation of agricultural policy through greater participation by state and local officials" (p. xv). While Uwadibie notes many positive consequences of the creation of states through his analysis of Nigeria's agricultural policies in this context, he seems to arrive at the same conclusion as Yesufu. State creation has not promoted development though he stops short of concluding that it has increased underdevelopment.

The second on Yesufu's list of the problems with underdevelopment in Nigeria is the culture of consumerism and waste that enables conspicuous consumption rather than savings and investment. These practices are

> manifest in lavish parties for, and on the occasions of acquisition of, meaningless chieftaincy titles, on marital and even burial occasions, child-naming, etc; heavy dowry payments in many parts of the country which young men are compelled to make against borrowed funds, and which are frittered away in drinks and other forms of conspicuous consumption...(and) overdressing by politicians, the *nouveaux riche* and the elite, who feel compelled to maintain a pompous public image... (Ibid.)

Yesufu's identification of consumption, rather than savings, is clearly based on the classical economic assumption that savings guarantee capital for investment in an economy. As discussed in Chapter 2, early modernization theorists such as Lerner (1958) argued that it was necessary for people in "traditional" societies to be exposed to Western ideas of savings and investment through the technologies of information and communication. This notion is also integrated in Rostow's second stage of development—the precondition for takeoff—which includes exposure to Western ideas through the media, restlessness and eagerness for change in the worldviews of traditional peoples. Lewis (1967) also stresses the importance of savings in generating capital for investment. In the Nigerian case, much of its current development problems, despite its abundant resources, are rooted not just in lack of capital to invest in productive ventures, but in overspending on consumer goods. Many of these expensive goods are imports that drain the economy of vital resources and therefore not good for sustainable development.

Thirdly, Yesufu attributes the problem of underdevelopment in Nigeria to the prevalence of bribery, corruption and embezzlement of public funds which "have come to constitute not just traits of trade, but virtually the most potent industry in Nigeria of the 1980s and 1990s (Yesufu, 1996: 4). This problem is compounded by the phenomenon of capital flight. Those who embezzle public funds do not invest the money in the country but deposit in foreign bank accounts or procure property in other countries thus generating economic growth and development in countries other than Nigeria. For instance, in 1990, a Nigerian newspaper, quoting a report of the Morgan Trust Guarantee Bank of New York, wrote about how "some foreign bank vaults are bursting with $33 billion stashed away by some wealthy Nigerians" (Cited in Ihonvbere, 1999: 133-134). At the time, these deposits were said to be the equivalent of Nigeria's foreign debts, and certainly money that was not being invested in the Nigerian economy.

Fourthly, insecurity of life and property, especially through armed robbery, kidnapping and assassination, which has increased as a result of the general poverty, desperation and lack of opportunity for growth in the country, discourages tourism and foreign direct investment. Yesufu adds that the high incidence of political instability in Nigeria, starting with the first military coup in January 1966 exactly six years after the country regained its political independence, a 30-month bloody civil war, followed by years of military regimes, coups and counter coups, also exacerbated the problem of underdevelopment. A democratically elected government has ruled Nigeria for just 20 of its 48 years of post-independence history. The current Yar'Adua Administration was inaugurated in May 2007 after eight years of the Olusegun Obasanjo administration. This was the first democratic alternation in the country in more than 40 years.

Such an unstable environment does not encourage serious multinational corporations interested in investing in sustainable ventures. Many foreign businesses in Nigeria are in the oil industry, and though oil is Nigeria's major income earner, activities in this sector are largely non-productive in terms of enabling manufacture of consumer products in the country. This sector does not, therefore, directly generate real economic growth, because an economy that is overly dependent on imports (especially of consumer goods) is not self-sustaining. Also, in the age of globalization and footloose multinational corporations (MNCs) seeking production sites abroad Nigeria continues to be bypassed even as other countries provide incentives to attract them. An Export Free Zone was established in Calabar, a south-eastern city in the 1990s. It was only in 2001 that there was any significant activity in the area: diversion of used vehicles imported from Europe to the Calabar seaport as an effort to decongest the two sea ports in Lagos.

Evolution of Economic Planning in Nigeria

Yesufu's insights capture a general consensus among scholars, development practitioners and Nigerian policymakers alike that Nigeria is a highly underdeveloped country. As he notes, while minimal economic growth has occurred in the country, it has been too slow to translate into development (1996). But the state of underdevelopment in Nigeria is not for lack of policy. Nigeria's leaders, at least since independence, have often formulated policies and executed projects and programs aimed at promoting development in the country. The country has particularly relied on economic planning as a development strategy. Okigbo (1989) defines national economic planning as:

> a scheme by the state for the deliberate and systematic manipulation by state organs of forces, economic and non-economic, for the control of the economic environment organized around a set of stated goals and objectives together with a specification of the means for achieving them within a defined time period through a rational use of national resources. (p. 37)

The first known national economic planning was the Goelro Plan of 1920 for the electrification of the Soviet Union. It was not until the Great Depression of the 1930s in the United States and the rise of Keynesianism, welfarism and the need for state intervention that economic planning began to shed some of its Communist identity. For countries emerging from colonialism, starting with India in 1947 with the creation of the Indian Planning Commission, national economic planning seemed to be a development accelerator. It was the platform for many development projects in various Asian countries and Japan after World War II. There are three types of planning: the Marxist type (imperative planning) common in the former Soviet Union, the former Communist countries of Eastern Europe and China; the French system of indicative planning; and forecast planning found in Scandinavia. In the first model, "instructions are given from higher to lower hierarchies. The reconciliation of the instructions and requests brings forward material balances which tend to be more and more difficult to manage" (Okigbo, 1989:151). In the French system, "plans are not strictly binding on the society" and there is full participation of the private sector. And in forecast planning the state informs the public of its intentions and its views on how the economy should go, as well as its expectations, policies and actions. As in the French system, the private sector fully participates in this model of national economic planning. The development history of Nigeria is littered with various development plans many of which may not perfectly fit the three classic types. It began with the colonial development plan.

Colonial Development

A colonial administration, by definition, promotes the interests and welfare of the metropolis rather than those of the colonized people. Any benefits that accrued from colonial programs to the colonized were unintended spill over that could not be avoided unless through "outright plunder" (Okigbo, 1989:8).[2] The British were therefore interested in harnessing and exploiting Nigeria's resources and any development projects, such as railway, were set up only with this objective. The British colonial administration in Nigeria, between 1900 and 1945, concentrated on fulfilling two of Adam Smith's duties of a sovereign—provision of internal justice, and protection from external aggression—without concerning itself with the last duty: undertaking of social works—income distribution, employment and stabilization. Inter-personal and spatial income distribution or employment did not feature on the agenda of the colonial administration since its priority was not to develop the colonial territory for its own sake. Colonial administrators conceived of the role of the public sector only as it facilitated the "exploitation of the resources for the benefit of the metropolis" (Okigbo, 1989: 8). In line with the objective of extracting resources from the territory, the colonial administration embarked on a two-phase economic development program.

In the first phase, 1900-1919, priorities were on the development of a transport network—ocean transport (between the metropolis and the colony and protectorate) and rail transport "for the development and exploitation of the agricultural and mineral resources in the hinterland" (Okigbo, 1989: 9; Williams, 1976: 18). Thus the Nigerian rail network, which has not changed much since colonial times, connected centres of agricultural produce and mining activities from the hinterland to the coast in the southwest. During the second phase of the program, controls on exports and imports were introduced and the production of local substitutions for goods previously imported was promoted. This was not an attempt to make Nigeria self-sufficient. Rather, it was a reaction to the depressed global economy, an outcome of the Second World War in a climate where most countries were yet to recover from the

Great Depression of the 1930s. At the end of the Second World War, with a new global emphasis on welfarism, the British Parliament under the newly elected Labor Party approved £200 million "'for the economic and social advancement' of its colonies; and at the request of the Colonial Office, the Nigerian Government formulated the country's first ever development plan which has as a specific objective, the welfare of its citizens" (Yesufu, 1996: 54; Rimmer, 1981). This was the beginning of planned development in Nigeria, with several national development plans—many of which were filled with more sloganeering than any real strategy for economic development (Yesufu, 1996).

A Model of Development Plan for Nigeria

In preparing Nigeria's first economic development plan, the bureaucrats contemplated the various models. Both the French and Soviet systems of planning required enormous skills and resources in data gathering and the Indian style (which provided a model for many developing countries) demanded a high degree of "discipline," according to Okigbo (1989) who argues that Nigeria was lacking in these areas. The country therefore adopted a middle position that required merely setting out a program of public expenditure and accompanying policies of the economy. This model called for a greater participation of the private sector but in the early years of independence this sector was largely invisible in Nigeria's economy. Jarmon (1988) points out that in 1962 and 1963, for instance,

> the private sector contributed only 3.7 percent to the net domestic product. Under the conditions of the colonial order, indigenous merchants were only a marginal element in the new market arrangement and lacked ownership and control over the wealth being created in this sector. The lack of accumulated wealth in the hands of an indigenous merchant class was an obstacle to internal development (p.23).

While the program of development in Nigeria's 1900-1919 Plan can be said to be a national economic planning, it was not until 1946 that the country formally had a national economic plan with the Ten-Year Plan for Development and Welfare. This was followed by what was described as the first real "national" economic planning, the 1962-968 Plan, extended to 1970 because of the Civil War. There have been five national development plans in Nigeria since 1960. Each of these will be discussed in turn but first the 1946-1956 plan.

Ten-Year Plan of Development and Welfare for Nigeria, 1946–1956

Objectives: The Ten-Year Plan was, in many ways, the first long-term development plan in Nigeria, and definitely the first to incorporate the welfare of the colonial people. The Plan was enabled by funds approved by the British Parliament for the development of the colonies. The prevailing view at the time was that the "responsibilities of the state toward its dependencies must be similar to those accepted at home...and the thinking...indicated the more fundamental elements in standards of living as targets (Rimmer, 1981: 33). The primary objectives were therefore improvements in water supplies, nutrition and health. The Plan also stressed education, transportation and communications (Dean, 1972:11). The uneven progress of Nigeria, further worsened by the retrenchment polices of the 1930s following the worldwide recession made it imperative that "physical facilities regarded as the minimum necessary for the general improvement of the country and its population" and reduction in rural-urban migration should be provided (Okigbo, 1989: 20).

Criticisms

The Plan was aimed at improving the general health and mental condition of the people, and did not incorporate any attempt at general industrial development nor did it express explicit intention that developments in these areas contribute to macroeconomic output. In fact, the Plan was clear that it was "not assumed, however, that Nigeria will become an industrial country as with its large population and area a great deal of its future must rest in agricultural development in its widest sense, and the improvement of village industries" (objectives of the plan quoted in Okigbo, 1989:21). As Okigbo points out, the colonial officials who prepared the plan did not see Nigeria as "a potentially great country from an economic point of view, but (saw the country) as that of very happy peasants who drew their sustenance mostly from agricultural and pastoral pursuits, leaving the tedious task of manufacturing to the mother country" (p.20). There is a sense therefore in which the colonial administration saw development consisting of just the essentials to facilitate colonial exploitation of resources. The strategy for achieving these objectives included road construction, improvement in rural and urban water supplies, telecommunication, health and agriculture (including livestock, fisheries and forestry).

Again, in setting the targets for achieving the planned goals, it was clear that the colonial administration was not interested in overall macroeconomic growth. For instance, in 1945, the total trunk telephone mileage in Nigeria was 7,248 miles. The Plan compared this to Scotland's 74,370 miles (in 1925, rising to 296,623 miles in 1938) and the United Kingdom's 769,621 miles in 1925 (rising to 3,763,064 in 1938). While the document acknowledged these vast differences, there was no plan to significantly raise the mileage in Nigeria, proposing only to "establish 2,500 miles of new route, 2,300 miles of new wire on existing routes and 1,000 miles of reconstruction in the seven zones into which the country had been subdivided" (Okigbo, 1989:25). The disinterested attitude toward any real development in the country was also evident in the health sector where the colonial administration planned to provide one hospital bed for every 2,000 people as against one per 250 in the UK (excluding private facilities and specialist hospitals) during the same period.

Post-independence National Development Plan, 1962–1968

Objectives: This plan was considered truly national for two reasons: it was the first Plan in post-independence Nigeria and had a higher level of participation by Nigerians than the previous plan. Its objectives included the achievement and maintenance of "the highest possible rate of increase in the standard of living and the creation of the necessary conditions...that will be required" (Okigbo, 1989: 41). It prioritized agriculture, industries and manpower development, stressing food production although its main focus seemed to be on the production of export crops. It also aimed at pursuing development goals, conceptualizing development in similar terms as the previous plan: provision of water, clinics and health facilities, roads (particularly rural feeder roads) and primary and secondary schools.

As well, the plan adopted a pragmatic approach to development. Accordingly, emphasis was placed on the provision of amenities to the rural areas; standard of living was measured visually by their presence or absence. This continues to be the case in Nigeria where location of "amenities" or "industries" has remained a highly politicized matter. This would explain the logic behind the location of the second Nigerian oil refinery in Kaduna in the non-oil producing Northern part of the country. The location may be irrational in economic terms, but it is politically pragmatic.

Criticisms

The Plan certainly attracted higher levels of participation by Nigerians than the previous one. However with few Nigerian bureaucrats at the top echelons of the civil service, there was still considerable colonial involvement in the preparation and implementation of the plan, with many of the planners being Americans and Britons affiliated with the Ford Foundation. Also, the Plan was not national enough because the central government had devolved too much power to the regional governments which participated marginally in the implementation of the plan.

Furthermore, the Plan failed to integrate the role of, or acknowledge the participation of, the private sector in the national economy—except by reference to revenue contributions. While different private-sector organizations were consulted occasionally during the preparation of the document, there was no organized forum for the sector to "air its aspirations, except by means of periodic delegations to the minister concerned" (Okigbo, 1989: 39). As well, there was no feedback from the grassroots—especially farming communities.

More significantly, the Plan was neither rigorous nor articulate because it was formulated with insufficient facts (Dean, 1972: 29). The lack of "statistical and other basic information normally required" for preparing a development plan was to be the basis of a book by Stolper, the economist who headed Nigeria's Economic Planning Unit (under the auspices of the Ford Foundation) in the Federal Ministry of Economic Development that prepared the 1962–1968 National Development Plan (Adebo, 1966: xii). Also, from the perspective of promoting industrialization, the "planning machinery was not equipped to deal adequately with detailed industrial planning as distinct from broad economic planning" (Onyemelukwe, 1966: 12).

Second National Development Plan, 1970–1974

Objectives: The first Plan after the Nigerian Civil War, 1967–1970, the Second National Development Plan accordingly focused on reconstruction. The government aimed at building a "united, strong and self-reliant nation, a great and dynamic economy, a just and egalitarian society, a land of bright and full opportunities for all citizens, and a free and democratic society" (Okigbo, 1989: 79). It emphasised the creation and growth of resources as a mechanism for solving the problem of inequality, much of which was cited as one of the causes of the Civil War. This Plan, like the ones before it, also stressed the welfare of the individual through equitable distribution of resources. It reaffirmed the priority areas set out in the 1962-68 Plan: agriculture, industry and the development of high-level and intermediate-level manpower. Its centrepiece was indigenization of businesses through the promulgation in 1972 of the Nigerian Enterprises Promotion Decree (also known as the indigenization decree) which "brought a measure of indigenous control over the private sector" (Jarmon, 1988: 24). The decree reserved specific economic opportunities for Nigerians and required Nigerian participation in firms engaged in a wide range of other activities (Williams, 1976: 47). The Plan, executed in the oil boom era, was more successful than any other plan before or after it. Its most conspicuous achievement was in the education sector. Primary school enrolment rose by 28.6% from 3.5 million in 1970 to 4.5 million in 1973, an average annual growth rate of 9%.

> Secondary enrolment also virtually doubled from 343,000 in 1970-71 to about 649,000 in 1973-74. Compared with a national population growth rate of 2.6%-3%, this represented a positive gain. But the average growth of education investment over the same period, 1970-75 was 6.8%. This suggests a fall in the quality of the education available to the increased pupil enrolment (Yesufu, 1996: 70).

Criticisms

In his analysis, Okigbo (1989) argues that the formulators of the Plan proceeded as if the Civil War had not happened by ignoring the creative innovations of the people of Biafra.[3] All the productive capacity developed for the war was never integrated into national development, and therefore policymakers failed to tap into its potential to contribute to industrialization.

The Plan was later revised to increase government investments due to the unexpected flow of revenue from petroleum. However the petro-dollar created its own problems. For instance, two major policy instruments of the plan were the stabilization of prices and the reduction of inequalities of income distribution. But the increased revenue prompted government to raise wages, unintentionally pushing the rates of inflation to double digits. A liberalization of imports "which created unfavourable competition to domestic producers" further exacerbated the inequalities in distribution (Okigbo, 1989: 99).

The Third National Development Plan, 1975–1980

Objectives: Although the 1975–1980 Plan reaffirmed the objectives of the previous plan, it advanced a more ambitious agenda. These included: an increase in per capita income, a more even distribution of income, a reduction in the level of unemployment, balanced (geographically dispersed) development and an increase in the supply of high-level manpower (stress on education). Others were a diversification of the economy, balanced development, and indigenisation of economic activity. The Plan incorporated, for the first time, a deliberate effort at developing Nigeria's communication infrastructure. About 5.1% of annual budgets during the Plan period were allocated to the telecommunications sector. This represented ₦1.338 billion over five years. As at 1974-75, Nigeria had one phone per 667 inhabitants and one postal service for 42,000 people. This represented an increase from 70,000 to 109,000 telephone lines during the Second Development Plan, 1970-74 (14% per annum). A ten-fold increase in targets was set for the 1975-1980 Plan such that at the end of the period the number of exchange lines was expected to increase from 52,000 to 500,000, and telephone lines from 109,000 to 1,000,000. The number of telex lines was expected to climb from 594 to 6200. Also, there was a proposal for "219 new radio and line carrier routes, expansion of microwave radio system, coaxial cable transmission between Lagos and state capitals, additional power plants as standby power for the exchanges and transmission terminal stations; high-frequency radio links between Lagos and state capitals" (Okigbo, 1989: 122).

These targets were achievable with deployment of the funds allocated to the sector. "However, since practically every item would have had to be imported and then installed by resident companies, the importance of bulk handling and port facilities could not be neglected. On paper, therefore, the targets expressed the yearnings of the people rather than practical reality" (Okigbo, 1989: 123). In the end, the goals were not reached though there was a significant increase in Nigeria's teledensity during the period.

The main strategy for achieving the general goals of the Third National Development Plan was the utilization of oil revenues in the development of the productive capacity of the economy, control of inflation and a guarantee of a more egalitarian income distribution. As 70% of Nigerians lived in the rural areas, the Plan prioritized programs and projects that would directly benefit the rural population and reduce the income gap between them and urban dwellers. Such projects included the importation of consumer items through the government-owned National Supply Company (and sold to the public at reduced prices), payment of higher prices for farm produce by the marketing boards, subsidies on agricultural inputs—such as chemicals, fertilizers and seeds—price control on consumer items, rent control and "a wage freeze while improving the remuneration of lower grade wage earners," investment in education, support for small-scale enterprises and use of local materials, labor and skills (Okigbo, 1989: 106).

Criticisms

While these programs were executed with good intentions, they only succeeded in widening the income gap, especially between urban and rural populations. Usually, the major recipients of the welfare programs embedded in the Plan were the middle class and higher-income groups. In the colonial days, the marketing boards were used to subsidize the British consumer, and to shore up the reserves and balance of payments of the sterling bloc. In neo-colonial Nigeria, indigenous rulers used the marketing boards to finance party political activities and enrich themselves.

As Lipton (1976) and Bates (1988) have also argued, programs aimed at redistributing wealth by raising income through subsidies on agricultural inputs always had an inherent urban bias. In many cases, those who benefited from these subsidies were the wealthy, large-scale urban farmers, or some middlemen, who then re-sold the products at huge profits to the poor and small-scale rural farmer. In controlling the prices at which farm products were sold to the marketing boards, the Nigeria government—as well as many African governments—only succeeded in making farm production an expensive and unprofitable activity. This then exacerbated the problem of rural-urban migration and increased dependence on imported food products. The importation of consumer items, as Okigbo points out, also discouraged local production, thus defeating government's intention of increasing self-reliance and the capacity of the country to feed its people.

The Fourth National Development Plan, 1981–1985

The Fourth National Development Plan coincided with the return of civilian administration in the country (1979) and overlapped with the resurgence of military regimes (1983). It was buffeted, *ab initio,* by both the political crisis in the country (a military coup and change in leadership and regime structure) and declining prices of petroleum products, the major source of Nigeria's foreign exchange earnings (which financed imports as well as domestic expenditures). The 1980s are usually referred to as the "lost decade" for developing countries. It was even more so for Nigeria, rudely awakened from the oil-generated wealth of the 1970s to the oil bust, debts, stagnated economy and military dictatorships. Nevertheless, the fourth development plan started off with the same optimism that attended previous plans particularly as it announced a continuation of the Third Plan, with a few additions: greater self-reliance, development of technology, increased productivity, reduction in rural-urban migration, promotion of a new national orientation conducive to greater discipline, better work ethics and cleaner environment. The Plan redefined the concept of development to mean the "development of man – the realization of his creative potential, enabling him to improve his material conditions of living through the use of resources available to him" (Okigbo, 1989:134-135). It was anchored on three major objectives: economic growth and development, price stability and social equity. The machineries to achieve them were fiscal policy, monetary policy and income policy (Yesufu, 1996:78).

Criticisms

In the five years of the Plan, the country recorded negative growth rates (Okigbo, 1989: 147). It failed to stabilize prices as declining per capita gross domestic products and inflation "resulted in rocketing consumer prices" (Yesufu, 1996: 85). To add to the declining income levels,

> the cost of living was simultaneously escalating very fast. Food prices rose by an average of 54.9 % per annum; the lowest rate of increase of 12.2% per annum was recorded with respect to accommodation partly because of the rigidity and the difficulty of getting tenants to agree easily to rent increases. The composite consumer index rose by 95.5 % over the period 1981-1985, thus recording an average yearly increase of 23.8% (Yesufu, 1996: 85).

The Structural Adjustment Program Years, 1986–1988

There was a three-year gap between the fourth and fifth development plans for several reasons. The most obvious was the military coup of 1985 and yet another change in political leadership. Also, the economy had stagnated to the extent that when the Babangida Administration came in, it took time to decide on the best path to economic recovery, eventually choosing Structural Adjustment Program (SAP), as prescribed by the International Monetary Fund—suggested initially as a quick fix to revive Nigeria's economy. There is vast literature on SAP in the context of developing countries generally and particularly in Nigeria (see for instance, Anunobi, 1992; Haggard and Kaufman, 1992; Ihonvbere, 1994). However, an overview of key issues surrounding the implementation of SAP is necessary in order to understand the context in which the fifth national development plan unfolded.

After a Civil War and nine years of military rule, Nigeria's Second Republic began in October 1979 with Shehu Shagari as president. The four years of this administration coincided with the oil bust as global oil prices plummeted. The corruption of the politicians did not do much to improve the situation.

The Nigerian economy during the Shagari years (and right up to 1985) was characterized by incessant corruption, misappropriation of public funds, capital flight, inflated contracts, ghost workers[4] and general financial indiscipline. Public officials appropriated public resources as their personal entitlements, acquired private property abroad with public funds and led lives of appalling wasteful consumption, especially of imported consumer items, thus crippling local industries. The politicians' vulgar opulence thrived alongside abject pauperization of the majority of Nigerians, with the elected officials seemingly oblivious to the suffering of the people. A federal minister, whose ministry was responsible for the importation of rice, was reported widely as telling journalists that it was not true that Nigerians were starving as he had not yet seen anybody eating from the dustbin.

Due to the scarcity of foreign exchange, a new market in import licences soon emerged. People with connections to political power would apply for and receive import licences ostensibly for the importation of raw materials for local factories. An import licence facilitated access to foreign exchange at the Central Bank of Nigeria—at the official exchange rates. Those who used their political connections to obtain the license would then re-sell it at exorbitant prices to actual manufacturers who desperately needed the licences and the foreign currency to procure raw materials for their factories. This was an era of reckless influence peddling and name-dropping and people became instant millionaires just by being associated with those in political power. The corruption of the politicians in the Shagari Administration and the visible opulence of this class in the midst of such desolation prepared the stage for the military to return to politics in 1983.

The Buhari Administration

The duo of General Muhammadu Buhari and Colonel Tunde Idiagbon came in through a successful coup d'état on New Year's Eve, 1983. They began to tackle the nation's crisis through stiff austerity measures. They attempted to inculcate social discipline and good work ethic in Nigerians through programs such as the War Against Indiscipline (WAI). The inflationary trends were tackled with price control measures and sale of "essential commodities" to people at government supply stores.[5] Between 1984 and 1985, the economy recovered slightly. For instance, the GDP per capita in 1985 was ₦908, up from ₦852 for each of the previous two years. The government made debt repayment a high priority with a significant percentage of the country's foreign exchange earnings allocated accordingly. It considered the proposals for economic recovery offered by the IMF but rejected them in favor of an alternative framework. This consisted mostly in exploring the possibilities of exchanging crude oil for essential imports with some willing South countries; as well as enforcing domestic financial discipline especially in reducing consumption and corruption.

But it was not long before Nigerians got tired of "all that discipline" particularly when the media began to report contradictions between the speeches and professed intentions of policymakers and their actions. For instance, in the infamous "Suitcases Incident," some people from Sokoto reportedly slipped several suitcases filled with cash into the country. This followed the change in currency in 1984 aimed at stamping out "corruption, counterfeit currency, and the black market by neutralizing money held outside the country and forcing hoarders to account for their cash" (Forrest, 1995: 99). To achieve the objective, "individuals were allowed to exchange only a maximum of ₦5,000 in the bank unless they swore an affidavit as to how they acquired the excess" (Ibid).

Someone (or a group of persons) with lots of externally hoarded cash in the Nigerian currency about to be phased out brought the money into the country to exchange for the new notes. Nigerians expected that the money would be confiscated and the alleged offenders arraigned before the appropriate authorities. The matter was, instead, hushed up because, apparently, the culprit was an Emir (a politically significant traditional ruler) and therefore untouchable. Beyond these obvious contradictions between words and actions, the repressive nature of the regime, which used decrees to suppress press freedom and personal and political liberties, was enough to signal the beginning of its end. Most notable of these decrees were Decrees 2 and 4 of 1984. The first bypassed the legal process of *habeas corpus* by legalizing the indefinite detention of persons without trial while the second made it criminal for journalists to write any story considered a falsehood or embarrassing to a federal government or public official. The military government arbitrarily determined what was embarrassing and false in a news story.

In the end, the administration's "ethical campaign embodied in the War Against Indiscipline (WAI); counter trade with Brazil, Malaysia, Austria and Italy involving the bartering of Nigerian oil for essential imports; and the repression of all opposition, especially intellectuals, students, and workers, alienated the regime from the populace" (Ihonvbere, 1994:118). It was not long therefore before Nigerians began to get restless for another change. This came in August 1985 in the form of another military coup that ushered in the eight-year administration of Ibrahim Babangida. This administration was more willing—and some would say, eager—to accept IMF's economic prescription for Nigeria's ailing economy.

The Babangida Years and the Structural Adjustment Program

When Babangida became Nigeria's first military president in 1985, the economy was virtually in shambles despite the spirited efforts of Buhari and Idiagbon to introduce some fiscal discipline. The Babangida Administration considered the prospects of taking an IMF-guaranteed loan and introducing the attendant structural adjustment program—a package that had been rejected by Buhari in favor of alternative measures such as counter trade. But in a tactical move to gain popularity and thus legitimize his authority, Babangida "referred" the matter to the people. However the questions raised in the debate were structured such that the emphasis was on national pride, rather than on the more fundamental economic issues of currency devaluation and the extent to which it would be executed. The infringement on national sovereignty was paramount in the minds of the people as they engaged in the "IMF Debate." As Yesufu (1996) notes, a genuine attempt to arrive at an informed decision would have involved consultation with those who really understood the economic issues at stake, and not thrown to people most of whom were hearing about the IMF for the first time during the debate and probably did not understand what devaluation of the currency entailed.

The questions were also framed such that what was clear to the people was that they were willing to make the sacrifices on their own terms rather than be further indebted to foreigners. As Forrest (1995: 210) points out, the debate centred "on the issue of the loan and not on the policy conditionalities involved in an IMF package or the wider questions of economic strategy." And thus structured, the IMF debate already had a predetermined response by Nigerians the majority of whom chose to reject the loan but make the sacrifices toward the recovery of their national economy.[6]

The sacrifices that Nigerians signed on to were the infamous "IMF Conditionalities" and involved the devaluation of the Naira, rationalization of the civil service (retrenchment of workers), market liberalization especially to attract foreign investments, abolition of price controls, scrapping of the marketing boards and removal of subsidies on services and goods (such as petroleum products). Prior to the adoption of the SAP conditionalities, on October 1, 1985 in his Independence speech, Babangida had announced a 15-month National Economic Recovery Plan, and an economic package that "closely followed IMF and World Bank thinking" (Forrest, 1995:211). But it was not until July 1986 that the structural adjustment program was formally introduced though some of the conditionalities had already been quietly implemented. The SAP "marked a shift in economic policy towards a strategy that relied more on market forces and private enterprise to promote national accumulation" (Ibid., p.207).

The immediate effect of SAP was soaring prices of consumer items beginning only hours after the September 1986 devaluation of the Naira by 69%, closing at ₦5.06 to the US dollar at the end of the first auction at the Second Tier Foreign Exchange Market (SFEM).[7] The effect of the devaluation on the price of consumer items was drastic, and it was not only on non-essential items: food and transportation costs rose, literally overnight. Soon the cost of SAP was more than the people had bargained for.[8] For one thing, the package had been sold to Nigerians as a sharp but quick pain, "designed to rapidly and effectively transform the national economy over a period of 15 months and to end by June 1986" (Yesufu, 1996: 91). Three years later, in 1989, the people, especially in the South, took to the streets to protest the pains of SAP. Even the government had to admit that SAP had been more painful than had been anticipated, but continued to justify its necessity (and its gains) in a Federal Ministry of Information publication.

The Fifth National Development Plan, 1988–1992

Objectives: It was in this economic environment that the Fifth National Development Plan, 1988-92, was launched. It basically affirmed the objectives of SAP. The premise of the Plan was to open the Nigerian economy to greater domestic and international competitiveness in order to create a free market for goods and services necessary for growth.

> The Fifth Plan dealt with short, medium and long term objectives which support the objectives of SAP—achieving fiscal balance and reducing the balance of payments deficits; laying the foundation for non-inflationary growth; restructuring and diversifying the economic base of the economy to reduce dependence on oil (Aka, 1999: 35-36).

The Plan stressed efficiency and equitable redistribution of resources targeting the gap between urban and rural dwellers, with policies aimed at addressing the conditions of the latter. A key instrument to achieve this goal was the creation of the Directorate for Food, Roads and Rural Infrastructures (DFRRI), which aimed specifically at creating infrastructures such as roads in the rural areas. It was understood that the problem of food scarcity, for instance, was not so much that of inadequate production but of poor distribution and storage. Certain infrastructures such as roads and electricity were therefore needed to address this shortcoming.

However, it was widely argued that DFFRI only succeeded in opening up the rural areas—in the few places where the program succeeded—to exploitation by urban dwellers. In Lagos state, farmers and fisher folks from the surrounding rural areas and their produce were weekly transported to "Sunday markets" in metropolitan Lagos. The food items were sold to urban dwellers at relatively lower prices. The justification was that everyone benefited: the rural dwellers (mostly women) found a market for their produce and earned more returns than they would have if they had sold their produce only in their communities; the urban dwellers (also mostly women) benefited from access to less expensive produce. If this had eliminated the activities of middle persons, perhaps it would have been a positive sum transaction. As it was, not all rural farmers could be transported by the state government to city markets. Market women from the cities, benefiting from the easy access to the rural areas (through feeder roads) would go into the rural areas and buy agricultural items—such as vegetables, yams, snails and garri[9]—at very low prices and transport to city markets where they would re-sell the products for enormous profits. In the end, the middle persons (in this case market women from the city) gained more from the opening up of roads than the rural dweller who still battled the problem of storage and therefore succumbed to the enterprising and merciless (in terms of driving down prices) city woman.

If small-scale producers of food crops did not significantly prosper during this period, the large-scale cash crop producer benefited immensely from the scrapping of the marketing boards—initiated in the early years of SAP. With easy access to the international market and removal of price controls, producers of export crops such as cocoa relied on the market forces of demand and supply to determine the prices of their products. From anecdotal accounts, the market was good to them. In southwestern Nigeria, where most of the cocoa in the country is produced, there emerged overnight rich men, and many of them celebrated their new wealth by marrying second or third wives and acquiring chieftaincy titles.

In such an environment and in tune with the economic mood of the period, the Fifth Plan, unlike previous ones, stressed the role of the private sector particularly in its capacity to create employment opportunities through the combination of an integrated rural development program and the establishment of small and medium scale industries (Okigbo, 1989:187). This was not surprising given that this was the age of SAP and external-oriented strategies of development. Market forces accordingly replaced administrative controls.

Criticisms

In reviewing the Plan (and the previous ones), Yesufu (1996) notes that it failed to meet its targets with many of the objectives sounding like slogans and efforts to "please every citizen and group that would be articulate enough to read the plan or to ask questions" (p.63). As Wright (1998) notes, the four plans before SAP proposed mixed development, combining strong government participation with private industry such that the role of the government was always critical. But they "failed to reach their ambitious goals, although certainly some limited gains were achieved" (p. 118).

Post-Fifth National Development Plan

The Babangida Administration that supervised the Fifth National Development Plan ended in chaos in August 1993. Two months earlier, it had annulled the results of the first presidential election in ten years, throwing the country into a political turmoil that arguably persists even as this book is being concluded. In August, Babangida was forced to "step aside" and a transitional government, headed by a businessman, Ernest Shonekan, was appointed and tasked with organizing fresh presidential elections. In November, Sani Abacha, then the defence minister, announced Shonekan's "resignation." The decree that had set up the Interim National Government included the clause that in the event of the resignation of the head of government, the defence minister would take over. And thus began the Abacha years which ended abruptly with his death in June 1998. This period was one of the most politically turbulent and repressive in peacetime post-independence Nigeria. At the economic front, while there was a National Economic Planning Commission—with a foremost economist, Sam Aluko, at the head—the Abacha administration did not pay much attention to economic planning. And it was not long before the commission was abolished. "Abacha apparently did not have the intellectual readiness for such a joke" (Omo-Ettu, 2007).[10] In place of a national development plan, the Abacha government introduced what it called Nigeria's Vision 2010.

Abacha and Vision 2010

In September 1996, Abacha announced his Vision 2010 committee initially composed of 170 members but later expanded to 250. Shonekan, the man whom Abacha overthrew only three years earlier, chaired the committee, tasked with considering the best-suited economic development program for the country. In Abacha's words, the committee was expected to "define for our nation its correct bearing and realistic sense of direction" (cited in Wright, 1998: 123). The committee was composed of government ministers, academics, journalists, traditional rulers, trade union leaders and foreign businessmen, and representatives from the private sector. Vision 2010 had four central prongs: democratization, liberalization, globalization and technology.

When the committee first met, they identified 13 critical factors under four broad categories:

> human capital (health, education, population); shared values (norms and standards, anti-corruption, openness, cooperation and managing diversity); governing systems (law and order, good and stable governance); global competitiveness (external environment; science, engineering and technology, competition, sustainable economic growth).[11]

The Vision 2010 committee submitted its report in September 1997 recommending a "'large-scale deregulation of the Nigerian economy' (and) the release of political detainees and rigorous compliance with the transition program" (Human Rights Watch, 1997).

> In his October 1, 1997 National Day address, Abacha promised to "introduce the measures immediately required to begin the program's implementation in the firm belief that succeeding administrations will carry it to a successful conclusion with the support of all our people and friends of the nation" (Ibid).

There were doubts even among committee members that the Abacha government would implement any of the recommendations. In any case, the work of the committee was criticized for focusing on agriculture and industry in an era when the rest of the developing world—namely the Asian Tigers and India— were developing and generating revenues from its information technology industries. A Nigerian-born American computer scientist, Philip Emeagwali, said Vision 2010 was modelled after Malaysia's Vision 2020, and yet it did not have the long-term goal that the Asian model had. He argued that if "Vision 2010's goal is for Nigeria to derive its entire wealth from agriculture and industry," it would turn Nigerians into "the hewers of wood and fetchers of water for those nations that have arrived in the Information Age" (Emeagwali, 1997).

Nigerian Economic Policy: 1999 to the Present

After 15 years of military rule and political and economic turbulence, General Obasanjo again assumed political leadership of Nigeria—this time through the ballot box as a civilian president. Thus, Nigeria's Fourth Republic (and as many number of attempts at democratic governance) began on May 29, 1999.[12] In his inaugural speech, Obasanjo made only a passing reference to the economy. In December 1999, the seven-month old administration released a four-year national economy policy. The government did not call this program a national development plan. The 1999–2003 policy was guided by the following principles:

- The economy exists for and belongs to the people, and at all times the general well-being of all the people shall be the overriding objective of the government and the proper measure of performance.
- Given the state of the economy which is equivalent to national emergency, economic management shall involve total commitment of the leadership at all tiers of government and the mobilization of the populace without creating a bloated government.
- Government shall be lean, efficient, honest, transparent, cooperative and friendly, operate on the basis of extensive devolution of power; and shall function mainly as a facilitator.
- Government's primary role shall be to ensure, in cooperation with the private sector, the urgent creation of adequate and efficient infrastructure, particularly of energy, telecommunications, water and financial services, to bring about a positive and internationally-competitive environment for economic activities.
- Private enterprise, private effort, and non-governmental action shall play the major role in achieving the goals of the society and the derived targets of the government.
- Everything shall be done to foster a strong work ethic to drive productivity (Federal Government of Nigeria, 1999).

Objectives and Instruments

These principles meant that the new Nigerian economy would be market-oriented, private sector-led, highly competitive (internally and globally), technology-driven; broad-based, humane, open and internationally relevant. The policy was targeted at achieving economic revitalization and growth, a significant rise in the standard of living, employment creation, participation in the global economy through a restructuring of the Nigerian economy to position Nigeria as the centre of economy in the West African region.

On page 8 of the 11-page document is a section on "Information and Communication Technology" declaring the following:

> Government will create incentives to expand access to information and communications technology which will facilitate leapfrogging in order to short-circuit the longer span of development. Government will encourage local production of ICT equipment and materials (computers, telephones, TVs, etc.). Government will also encour-

age the development of payment systems which will facilitate the growth of electronic-commerce (Federal Government of Nigeria, 1999: 8).[13]

It is not surprising that this document gave such scant attention to the development and diffusion of ICTs. The Obasanjo Administration did not prioritize this sector early in its administration, conceptualizing of ICTs as "newfangled notions of globalization and information technology" (Chiahemen, 1999). Ironically, the "revolution" in the sector occurred during that administration. Huge strides were made—through ICT-related policies that enabled deregulation of the telecommunications sector and licensing of private telecommunications operators—during the Obasanjo government. He himself ended up being a reluctant convert to the "new fangled notions" of ICTs.

Finally, it must be noted that the fifth national development plan which, like its predecessors, did not achieve its targets before the expiration date remained the last articulated national development plan in Nigeria. In August 2008, Ayodeji Omotosho, the coordinator of the economic analysis unit of the Nigerian Planning Commission, said a new economic plan would be announced in October 2008 (though it was yet to happen, as at October 15). According to newspaper reports, the plan is the:

> outcome of "harmonization" of National Economic Empowerment and Development Strategy (NEEDS-2) and Seven-point Agenda (of the Yar'Adua Administration). The new Plan is expected to run in three trenches up to the magical year 2020 when Nigeria is envisioned to be part of the 20 leading economies (Aremu, 2008).

ICTs and National Development Plans

From the foregoing review of Nigeria's efforts with the problems of underdevelopment, it is obvious that telecommunication was given some attention in only three of the development plans. The first was the 1900-19 which was not exactly a development plan because there was no intention to "develop" the country in any sustained manner. Communication was included only as it facilitated the administration of the colony. Telecommunications assumed importance again in the 1970-74 Plan, not necessarily as a development tool, but in line with the modernization mood of the period. The telephone, for example, was considered not so much as a development tool as a symbol of modernization. Attempts to develop the telecommunication sector began in the Second National Development Plan and continued into the third—1975 to 1980—though many of the targets were not met. While the Babanginda Administration did address this sector through the promulgation of Decree No. 75 of 1992 which set up the Nigerian Communications Commission, not much occurred until the Abacha years when for the first time in Nigeria, private operators were licensed to provide telephone services.

In Obasanjo's national economic policy, the development of ICTs was not a high-priority issue. However, the private sector pushed the issue to the top of the agenda of the federal and state governments such that the ICT4D became a national discourse. As a result, the development of the ICT sector emerged as one of the highlights of the Obasanjo Administration as it handed over to the Yar'Adua government in May 2007.

Conclusion

The state has always been active in Nigeria's development processes. The private sector was mostly ignored until the 1980s when it began to assume a more central role. This occurred in response to a variety of intersecting factors. For instance, following the debt crisis of the period, plummeting price of oil and the stagnation of the Nigerian economy, the government opted for economic internationalization, free trade, privatization and deregulation as packaged by the IMF in the structural adjustment program (SAP). This provided room for a more robust participation by the private sector, particularly with the privatization of hitherto government-run enterprises. However, the role of the state remains crucial in providing the regulatory framework for the economy generally and specifically for the development of the ICT sector. This point will be discussed further in the next chapter.

Development priorities constantly shift in response to domestic and/or external factors. For instance, the wave of global welfarism shortly after World War II undergirded the Ten-Year Plan for Development and Welfare (1946–1956) in which the colonial government attempted to incorporate the welfare of the colonial people by executing projects on education, transportation and communications. The Civil War not only interrupted the implementation of the First National Development Plan of 1962–1968, but also informed the direction of the second plan (1970–1974): infrastructural, economic and political reconstruction. While the different shifts from one plan to another responded to emerging realities and perceived needs, they also created a discontinuity in priorities, leading in many cases, to abandonment of projects. The result is that the same problems—such as poverty and unemployment—that previous plans aimed at solving persist, and constitute the content of current discourse on the linkages between ICTs and socio-economic development. It would therefore appear that the more things change (the shifting policies and priorities), the more they remain the same (the concerns and problems are constant).

Notes

[1] Reference to Lagos as a city requires some clarification since there is actually no city of Lagos. What is often referred to by locals as "Lagos" is Lagos Island, one of the 22 local government areas that make up the state of Lagos. But due to the metropolitan nature of Lagos (the state) and its level of development, the entire state is often considered as one huge metropolis. Even the least developed of the local government areas in Lagos, such as Epe and Badagry are more infrastructurally developed than some state capitals in other parts of the country.

[2] It could also be argued that there was no difference between colonialism and "outright plunder" as colonialism was inherently a plunder of a people and their resources.

[3] This refers to the technologies of warfare which the Biafrans developed to fight a war in which they were clearly outmatched in terms of manpower and arms. But their innovativeness and resourcefulness is said to have made up for their lack in other areas.

[4] This refers to the situation where wages in the civil service are paid out to workers who do not exist. Some civil servants would prepare the payroll such that names of non-existent people are included as staff and actual workers would sign for and collect the salaries of these "ghost workers." The situation was very common during the Shagari years and abated when the Buhari Administration insisted that people should physically identify themselves before they could receive their pay cheque. This created its own problems, namely long queues at the banks and delays in civil servants getting their salaries at the end of the month.

[5] This measure did not help because Nigerians with "connections" to secure allocations of the commodities at official prices opened "supermarkets" through which they re-sold the items to the public at exorbitant prices. Often, they would hoard the items to create an artificial scarcity that would further increase the cost to the public.

[6] In the end, the Babangida Administration did take the loan – without publicity – thus compounding the sacrifices of the people. It borrowed US$4.2 billion over three years, ostensibly to fund the Structural Adjustment Program. Besides, by the time the debate had been opened to the public, many of the conditionalities were already being implemented, or in the process of implementation. Babanginda only threw the matter to the public when he could not get the consensus of members of the Armed Forces Ruling Council on some thorny issues such as petroleum subsidy and currency devaluation.

[7] By the summer of 2001–15 years later—the Naira was exchanging at 136 to the dollar in the unofficial currency exchange market. It exchanged for 120 in August 2008.

[8] The cost of SAP was variously referred to in Nigeria as the "pains" or "grains" of SAP in response to government's frequent speeches about the "gains of SAP."

[9] Garri is a staple food in many West African countries. It is made from cassava.

[10] Titi Omo-Ettu, formerly of the Ministry of Communications and partially responsible for preparing the sector's inputs into previous development plans, in response to an e-mail question on why there was no national development plan during the Abacha years.

[11] Much of the discussion in this section is drawn from the annual reports of Nigeria by the Human Rights Watch, located at: http://www.hrw.org/reports/1997/nigeria/Nigeria-08.htm#P529_132846

[12] The Third Republic would have started from 1993 if the presidential elections of June that year had not been annulled by the Babangida Administration. This period is usually referred to as the "botched" or "failed" Third Republic.

• CHAPTER FOUR •

A Journey to the Future: The Policy Framework

Introduction

Nigeria came from a virtual ICT unknown at the turn of the 21st century to become one of the greats in Africa—at least in telephone usage. There was less than one main telephone line for every 100 inhabitants in August 2001 (0.43%) and a cellular phone subscriber base of less than 100,000 for a population of 140 million. Seven years later, August 2008, there were more than 1.6 million connected fixed lines and 53 million cellular phone lines in the country raising the country's total teledensity to 38.09 (Nigerian Communication Commissions, 2008). This has been a revolution, according a principal actor in the ICT industry in Nigeria.

The rapid increase in Nigeria's telephone penetration can be attributed to various factors. However, the specific success of the deployment of digital mobile telephony in the country is a triumph of the unique strategy adopted by the national regulator, the Nigerian Communications Commission (NCC). The Commission, enabled by the National Policy on Telecommunications (NPT), simultaneously opened up the telecommunications sector to competition while also holding a tight rein over the process. It licensed private-sector interests to provide digital mobile services, but protected the industry by restricting the number of digital mobile licences (DMLs) in the first five years. It also established parameters of operations such that the local service providers who offered fixed and wireless fixed lines could not venture into the digital mobile phone market. The Commission allowed an overt investor-oriented tariff structure ostensibly to safeguard the profit levels as incentives for investing in the industry. This rewarded existing investors while attracting new ones. Customers protested the high tariffs at every stage, leading to an upsurge in consumer rights groups and consumer rights advocacy. Paradoxically, the high tariff structure (and therefore high returns to investors) facilitated the rapid rise of the industry as operators invested more in infrastructure development and expansion of services.

The developments in Nigeria's telecom sector in the years between 2001 and 2008 have indeed been both phenomenal and revolutionary given the state of the landscape only a few years back. It however took Nigeria a long time to begin the journey that brought it to this point. As in many countries, the first steps were taken by the private sector, especially in the application of computers for basic word-processing functions. The print media were among the first sectors to recognize the promise of computers for more complex functions beyond mere functionality as advanced typewriters. As early as 1992, many print media organizations had switched from cut-and-paste to desktop publishing. Traditional "typesetters" and compugraphers were re-trained in computer skills. By 1994, many of these organizations, particularly those in Lagos, were word-processing news stories with computers. During my visits to some of the newspapers in 2001, it was observed that many organizations had eliminated the positions of typesetters and typists altogether because reporters were now typing their own stories and submitting them electronically to line editors, who also doubled as copy editors, thus eliminating (or at least minimizing) the need for sub-editors and graphic artists. It has also become common to see many journalists—particularly those covering the information technology beat in Lagos—lugging around top-of-the-line laptop computers and cell phones as part of their professional equipment.[1] The banks computerized their operations right from the onset.[2] As at 2001, about 56 banks in the country had internet access, websites and offered basic online banking services. By 2008, the number of banks in the country had declined to 25 following reforms in the industry. Conversely, the level of ICT usage had risen making the banking industry one of the most ICT-penetrated in the country.

Initially, the rate of ICT penetration was slow because of the absence of clearly defined policies on ICTs. This was considered by ICT practitioners to be a major hindrance to sustainable and purposeful ICT development in the country.[3] Jimson Olufuye, an industry activist, recounts the failed efforts of private-sector interests to get the National Committee for the Acquisition of Computer and Electronic Technology (NACACET), a committee appointed by the Ministry of Science and Technology, to take ICT development "more seriously."

> Though the objective (of the committee) was to provide cheap and affordable computers to Nigerian schools, business communities and homes, nevertheless the move failed because there was no clear Nigerian government vision on IT as a strategic tool for development. The long and short of it was that there was no IT policy for Nigeria by the then Federal government of Nigeria. ... The impact of this zero-policy on the IT

domain has grossly contributed to the underdevelopment of not only the profession and industry but also the Nigerian nation as a whole (Olufuye, 2001).

Eventually, the government introduced two policies that would guide ICT development and utilization in the country. These are the National Policy on Telecommunications (NPT) of 2000 and the National Policy on Information Technology (NPIT) of 2001. The evidence suggests that though private-sector efforts had succeeded in improving ICT usage and awareness, it was only when the state prioritized ICTs through its policies and programs that greater awareness and usage ensued. The central role of government has two implications. First, the involvement of the state creates momentum in the short run, which may be lost with change leadership, with the often attendant change in national priorities. Moreover, agencies of state tend to respond to new trends in development and shift attention elsewhere. Second, the fact that the state does not completely abandon the ICT industry to the private sector ensures that, theoretically, national economic goals can be achieved because government can moderate the rent-seeking tendencies of the private sector.

The National Policy on Telecommunications (NPT)

Background: Deliberate efforts to transform the Nigerian telecommunications sector from the colonial structure inherited at Independence began in 1992 when the military government of Ibrahim Babangida enacted Decree No. 75 of 1992. The objectives of the decree were broad, interpreting communications to include the electronic media as well as telecommunication. As its strategy for achieving the various objectives, the decree provided for the creation of the Nigerian Communications Commission (NCC). Inaugurated in July 1993, the NCC was given the task to:

- Create a regulatory environment for the supply of telecommunication services and facilities;
- Facilitate entry of private entrepreneurs into the market; and
- Promote fair competition and efficient market conduct among all players in the industry (Federal Republic of Nigeria, 2000: 13).

In an interview in Abuja, the chief executive officer of the NCC, Ernest Ndukwe, said the commission was established primarily to liberalize the telecommunications industry. And on the basis of that mandate, it had accomplished much since its inauguration.

> We can say that today in Nigeria, there is a liberalized telecommunications environment and in various aspects, Nigeria is actually ahead of several other nations in Africa as far as this is concerned. Today we have competition in the delivery of telephone access activities in a number of cities in this country. ... The customer has a choice of who to go to get a telephone line – it could be a fixed line or wireless line. ... In the satellite communication area, Nigeria is one of the most liberalized countries in Africa. We have a number of companies providing satellite hubs and providing local satellite links between one part of the country and another.

For the first time, there were private telephone service providers besides the national carrier, the Nigerian Telecommunications Limited (Nitel), and the presence of payphones in some cities such as Lagos that did not bear the Nitel logo. The private telecommunications operators (PTOs) provided local services but also had the capacity to terminate calls such that people calling their customers from outside their operating zones could get connected through to their parties. This expanded teledensity in Lagos, where several of the providers had found lucrative market, especially in the new development areas outside Nitel coverage area. Many of these companies, to overcome the infrastructural constraints (or perhaps to leapfrog antiquated technologies), provided their services through Wireless Local Loop (WLL), using radio waves for transmission rather than cables.

With the sector beginning to awaken, the NCC then set in motion the process that resulted in the formulation of the National Policy on Telecommunications (NPT), announced in October 1999, amidst media fanfare, by Communications Minister Mohamed Arzika.[4] He told journalists that the policy was aimed at helping the country to "achieve a modernization and rapid expansion of the telecommunication networks and services to enhance national economic and social development (and as) a major means of integrating Nigeria into the globalized telecommunication environment" (Aragba-Akpore, 2000a).

In the preamble to the policy, its authors noted that the

> availability of an efficient, reliable and affordable telecommunications system is a key ingredient for promoting rapid socio-economic and political development of any nation as telecommunications is a vital engine of any economy; an essential infrastructure that promotes the development of other sectors such as agriculture, education, industry, health, banking, defence, transportation and tourism. It is indispensable in times of national emergency or natural disasters. It considerably reduces the risks and rigours of travel and rural-urban migration (Federal Republic of Nigeria, 2000: 10).

They also underscored the need for Nigeria to have a functional and efficient telecommunication system for effective participation and equal partnership in the "emerging global market" and relevance in the "new millennium and beyond" because "global telecommunications provide the opportunity for a country to share in the wave of science and technology developments, and the general economy in positive ways, that account for the remarkable economic growth in advanced countries and the newly industrialized countries" (Federal Republic of Nigeria, 2000: 10). The policy not only set out to improve the country's telecommunication system, but to reverse the colonial trend where telecommunication facilities were "geared towards discharging administrative functions rather than the provision of socio-economic development in the country" (Ibid).

The poor state of telecommunication in the country framed the objectives of the NPT. The policy outlined three-year short-term and five-year medium-term objectives because policy formulators believed that "the rapidly changing nature of technology in telecommunications makes it difficult to set long term policy objectives."[5] There were nine short-term objectives that included:

- Implementation of network development projects to ensure that the country met and exceeded the ITU recommended minimum teledensity of 1 telephone to 100 inhabitants. In practice, at least two million fixed lines and 1,200,000 mobile lines would be provided within 2 years;
- Promotion of widespread access to advanced communication technologies and services, in particular the internet and related capabilities;
- Development and enhancement of indigenous capacity in telecommunications technology;
- Divestment of government's interest in state-owned telecommunications companies, such as Nitel and its mobile phone partner, M-Tel;
- Promotion of competition to meet growing demand through the full liberalization of the telecommunications market;
- Rapid resolution of licensing problems in the most equitable and transparent manner.

Some of the medium-term objectives were:

- The provision of a new regulatory environment sufficiently flexible to cope with new technological development and international trend towards convergence;
- Accessibility of telecommunications facilities to all communities in the country;

- Encouragement of domestic production of telecommunications equipment in Nigeria, as well as development of related software and services;
- Establishment of and reaching aggressive targets for the installation of new fixed and mobile lines;
- Encouragement of the development of an information super-highway to enable Nigerians enjoy the benefits of globalization and convergence; and
- Creation of the enabling environment, including the provisions of incentives to attract investors and resources to achieve policy objectives.

To meet these objectives, the policy targeted four areas of development: mobile cellular communication, appointment of carrier organizations, internet and web-based services, equipment manufacture and software development. The NCC was expected to establish guidelines for private-sector participation, and was mandated to issue licences to companies to provide services in the areas of: installation and operation of public switched telephony, terminal and equipment, payphones; provision and operation of private network links employing cable, radio communications, or satellite within Nigeria; mobile telephony; and repair and maintenance of telecommunications facilities. As noted earlier, the NCC had already started to accomplish some of these tasks as part of the process of deregulation and privatization of the telecommunications sector that began with the enactment of Decree No. 75 of 1992. But the NPT revised some of these tasks, expanding the role of the Commission and now requiring more fairness, equity and transparency from its members. Up to that point, the commission

> basically behaved like an organ of Nigeria's military set up and believed it owed nobody any explanation for its actions and consequently it was never part of its credo to explain its intentions, plans and actions to stakeholders and invite public participation in the formulation of such plans (Usoro, 2000).

However, with the exit of the military regime and return to democratic dispensation, things were bound to change. As Madamkor (2001) said: "The political environment in Nigeria is now different with a democratic government in place and therefore arbitrariness and inexplicable inconsistency are deemed to belong to the past." The policy also gave the NCC additional responsibilities such that its regulatory powers and functions extended far beyond broadcast and telephony into information technology. The commission was no longer simply licensing telecom operators, assigning frequencies and administering the national telephone numbering plan; it was also licensing internet service provision, operation of cyber cafés and satellite communication systems.

Implementation of the NPT

Three months after the 1999 announcement of the NPT, the NCC began to accept bids for four licenses to provide mobile telephony in the country. This was the first of several attempts over two years. There was considerable confusion and frustration for investors particularly when licenses were issued and cancelled in what appeared to be a very arbitrary process. In the midst of this, there were questions about the kind of technology to adopt in delivering mobile telephony in the country. The split was apparently a choice between the Global System of Mobile Communication (GSM, the system used in Europe and most of Asia) and Code Division Multiple Access (CDMA, the system used mostly in North America). A third alternative, considered the best of both worlds, was the integration of the two technologies into one system, according to an agreement reached during the 2000 World Radio Conference in Istanbul, Turkey. This was expected to facilitate universal roaming of cellular phone service, thereby increasing access. In Nigeria, government officials were reluctant to take a position because, as Communications Minister Arzika insisted in a newspaper interview in July 2000, it was up to the service providers to decide what technology would be more profitable for them (Aragba-Akpore, 2000b). As at the time of the auctioning of the digital mobile licenses (DML) in 2001, more than 96% of those who had expressed interest in the Preliminary Information Memorandum had indicated a preference for GSM, according to the NCC. This technology was therefore eventually adopted as the digital mobile technology of choice for the country. In 2006, the NCC issued unified licences which now incorporated both the GSM and CDMA technologies in the country.

The auctioning of the initial DMLs was a four-stage process: invitation, pre-qualification, auction and grant stages. At the pre-qualification stage interested parties were required to pay a refundable deposit of US$20 million to the Chase Manhattan Bank in New York. Successful bidders would pay the final license price less this amount while those not selected to the final four would be refunded the deposit with applicable interests. The auctioning itself began with an initial reserved price of US$100 million. The pre-qualification deposit and high reserved price were set to discourage non-serious bidders from participating in the process, according to the NCC. At the end of bidding, the highest offer was US$285 million; it became the final price for each license. Successful bidders were allowed 14 days to raise the capital. Three,

MTN, Econet and Nitel, met the deadline but one did not, arguing that it was seeking further clarification from the NCC before making the financial commitment. The federal government generated US$855 million from the exercise and transferred it as part of revenue allocation to the states.

There was a public outcry against the high cost of the licenses and the fact that the federal government did not apply the revenue to the development of telecommunication infrastructure. The perception was that government officials were only interested in profiting from the transaction. The result was that the operators were anxious to recover the cost of their investments within the shortest possible time and therefore less likely to interested in providing quality and affordable services. This concern was to bear out when the operators rolled out months later with tariff regimes that made Nigerians scream some more—at least, going by the many press conferences and letters to the editor. Many interest groups threatened to boycott the providers, accusing them of being "foreign companies" who just wanted to get in and make money and get out of Nigeria. They also noted that in Zimbabwe and South Africa where two of the three companies also had operations, the tariffs were much lower. The Nigerian government was equally criticized for being more interested in making money off the operators than in ensuring that Nigerians were protected from "foreign exploitation." This sense was further accentuated when the NCC took out full-page newspaper adverts to defend the operators and their tariff regimes.

The action of the NCC was in line with the NPT which stipulated that in order to attract private sector participation in the telecommunications sector, "the tariff structure shall be market-driven, enabling service providers and operators to recover their investments over a reasonable period of time" (Federal Republic of Nigeria, 2000: 58). During a personal interview, Ndukwe, chief executive of the NCC, defended the newspaper advertisements saying that it was a wrong perception that the commission sheltered the GSM operators to the disadvantage of consumers.

> We are to protect all. We don't protect one and leave the other. (At) NCC, our job primarily is to bring investment into this country in a liberalized telecommunications environment. We need to create the atmosphere that will encourage private sector investment into the industry. We need the big companies...We also have a duty to make sure that consumers are also protected.

Besides protecting the "big guys" to encourage them to invest, a public official said the NCC actually encouraged the initial high tariffs fixed by the GSM operators. This had been "necessary" to discourage Nigerians from scrambling for GSM phones before all the problems had been sorted out. But critics such as Titi Omo-Ettu, a telecommunications engineer and a principal

A Journey to the Future: The Policy Framework

actor in the sector, argued that if ICTs are to promote socioeconomic development, then licensing and tariff regime should deliberately favor access for all citizens to affordable telecommunication infrastructure. This would require lower licensing fees as well as a tax system that was attractive to investors.[6] He said since the federal government generated so much revenue from the licensing process, it should have used it to develop the telecommunications sector, which would reduce the transaction costs for the operators such that they would charge lower tariffs for their services. Already, there was a dire need to expand the country's public switched telecommunication network (PSTN), failure of which had and would continue to congest the expanding GSM networks.

On August 9, 2001 two of the licensees, Econet and MTN, rolled out their networks with an estimated capacity of 100,000 lines each, targeting monthly revenue of one billion Naira.[7] Previous developments in the Nigerian telecommunication sector were eclipsed by the GSM launch, hailed as epochal in the history of telecommunication in Nigeria. By December 2001, the three GSM operators had started full service, and recorded about 120,000 users. This number continued to climb and passed the 50 million mark at the beginning of 2008. There are also many more providers than the initial four. The significance of this development needs to be put in perspective. In about 136 years of telecommunication presence in Nigeria (starting with the telegraph in 1861), there were less than 500,000 functional phone lines. Also, in the four years that the small private telecommunications operators (PTOs) had offered telephone services (fixed line, wireless fixed line and mobile phone services) they had provided less than 100,000 lines—and these only in Lagos and a couple of other big cities.

During the first five years, several PTOs entered the market using the CDMA technology to provide digital wireless services. Initially, it was agreed that these providers could use their technologies to offer digital mobile services but charge the lower fixed telephony rates. A few of the companies began to do so. However, many of the PTOs agitated to provide services already embedded in their technologies, namely full digital services in voice, data and video. This started the debate over a unified license structure which heightened as advances in technology made restriction on the type of service that providers could offer difficult to implement. As Emmanuel Ekuwem, the vice president of the Nigerian Internet Group, pointed out:

> Everything is carried on broadband as data. Broadband doesn't differentiate between voice, data and video. There is a convergence of voice, data and video. Licenses that carry voice can carry video and data so it's not possible to restrict operators who can carry voice from carrying data only. This results in a painful attempt to regulate what

you can't regulate. Unified license gives you a one-stop shop center. Unified is an omni-service license and doesn't restrict the service. This aids and facilitates less cumbersome work of regulation. VOIP has really revolutionized this unified debate.[8]

Also it became clear that the "original four" providers had become well established enough to render the exclusivity clause unnecessary. The stage was thus set for the emergence of more digital mobile service providers in the country, even though "new" companies were not being licensed because providers seeking the unified licences must:

- Already be operating;
- Have been providing not less than 20,000 lines;
- Have paid their interconnectivity fees – must not be indebted to anybody;
- Demonstrate willingness to have a national spread.

The four original DMLs were conditional on the provision of at least five percent of services in the rural areas. The PTOs were not required to do this since they were mostly regional service providers, but in the new structure, they were mandated to provide services in the rural areas. At the end of the process, eight more providers entered the market on the GSM and CDMA platforms. The deployment of two different technologies was expected to minimize many service problems that users were experiencing. Also, the multiplicity of actors was expected to increase competition thereby reducing the cost of access and usage. The entry of the new actors raised two major issues in the sector: interconnectivity and infrastructure sharing; and a national backbone for the country.

Interconnectivity and Infrastructure Sharing

Right from the beginning of the operation of the digital mobile services, there were questions concerning why the four original providers, MTN, Intercellular, Econet and Nitel (also the national carrier), were not sharing infrastructure. It turned out that they were not permitted by law to share infrastructure and to associate because of concerns that they might collectively have too much power and therefore "gang up" against the public interest. The companies, on their own, acted in ways that constrained interconnectivity between them hence artificially creating many of the service problems that subscribers constantly experienced. An interconnectivity agreement was a condition of the DMLs but some of the providers, especially the national carrier, would arbitrarily disconnect other providers such that users could not initiate calls from one network to another.

In the new phase of the telecommunication development, providers were allowed to share infrastructure such as base stations, pipes, ducts and power to reduce overhead costs. The NCC issued interconnect exchange licenses to some operators to help in resolving the challenges of interconnectivity in the industry. While operators were now more able to work cooperatively with each other, the NCC sought to be informed of such sharing arrangement so it could stop any anti-competitive activities before they started (Omo-Ettu, 2007).

National Backbone

As at July 2007, there were two national carriers in the country, Nitel and Globacom.[9] Globacom competed for and obtained the licence to operate as the country's national carrier. Nitel and Globacom have national backbones—a country-wide transmission network. These national carriers are also operators and carry traffic as well as provide subscriber services. This distinguishes them from the other providers who are licensed to provide subscriber services but not to provide services to each other. Some companies in the country such as BCN are licensed to carry wholesale traffic—build the infrastructure and sell their services to providers without providing subscriber services. Nigeria can probably absorb more than two national backbones but the usual argument (mostly emerging from the exclusivity mindset of the first five years of telecom deregulation) is that the customer base is too small and spread out to support several backbones.

The National Policy for Information Technology (NPIT)

The second key policy on the Nigerian ICT sector is the National Policy for Information Technology (NPIT) of 2001 which specifically focuses on information technology (namely, the internet and computers—including the related hard/software aspect of the technologies). As its vision, the NPIT proposed to facilitate the process of making Nigeria "an IT capable country in Africa and a key player in the Information Society by the year 2005, using IT as the engine for sustainable development and global competitiveness."[10] Its specific mission was to use information technology for education, creation of wealth, poverty eradication, job creation and global competitiveness.

The National Policy on Telecommunications (NPT) had been widely criticized particularly for its narrow conception of telecommunication as telephone communication. This necessitated a national policy that would provide coverage for the information and communication sector while addressing the shortcomings of the NPT particularly in acknowledging the comprehensive nature of ICTs. The document, unveiled to the public in March 2001, was the result of consultations among interest groups in the industry culminating in a three-day national workshop on National Information and Communication Initiatives—Options, Policy and Plans in Abuja, March 2000. The workshop was conducted under the auspices of the Cooperative Information Network (COPINET) and sponsored by various organizations including the Nigerian government and Ford Foundation. The papers and comments presented at the workshop laid the framework for the NPIT. Relevant professional and trade associations such as the Computer Association of Nigeria (COAN), Information Technology Association of Nigeria (ITAN) and the Institute of Software Practitioners of Nigeria (ISPN) later submitted proposals to the policy drafting committee chaired by Gabriel Ajayi, a professor of telecommunications engineering at the Obafemi Awolowo University, Ife. He was subsequently appointed in June 2001 as the director-general of the Nigerian Information Technology Development Agency (NITDA) to coordinate the implementation of the IT policy. According to Ajayi, the need for a national policy on information technology became imperative after "the participation of the Nigerian delegation in the first African Development Forum on the Challenge to Africa of Globalization in the Information Age held in Addis Abba in October 1999."[11] After the Copinet workshop, "More efforts followed and culminated in the production of a master plan for the development of a national ICT program 'ICT 2000' during the term of Chief Ebitimi Banigo as Honourable Minister of Science and Technology" (Federal Republic of Nigeria, 2001: viii).

Given the extensive consultation with and input from various interest groups, the NPIT was well received when it was released. Many stakeholders from the private sector hailed the policy for being proactive and thus the tool needed to develop the ICT sector. This was to be expected because, as noted earlier, the document is mostly a product of the hopes and aspirations of the private sector. Ekuwem of the Nigerian Internet Group and a key participant in the process leading up to the formulation of the policy said:

> We wanted the implementation of the IT policy to be private-sector driven because the private sector is constant. They are investors. ... If the implementation of the policy is in the hands of private-sector organizations, then we are most sure ... of sustainability. The President has said that the new Nigerian economy should be market-oriented, private-sector-led and IT-driven. This is superb. This is the statement of Mr. President and we are all happy about that.

The NPIT sets out 31 general objectives some of which were:

A Journey to the Future: The Policy Framework 77

- To ensure that Information Technology resources are readily available to promote efficient national development;
- To guarantee that the country benefits maximally, and contributes meaningfully by providing the global solutions to the challenges of the Information Age;
- To empower Nigerians to participate in software and IT development; To encourage local production and manufacture of IT components in a competitive manner;
- To improve accessibility to public administration for all citizens, bringing transparency to government processes;
- To establish and develop IT infrastructure and maximize its use nationwide; To improve food production and food security; To improve healthcare delivery systems nationwide;
- To empower children, women and the disabled by providing special programs for the acquisition of IT skills;
- To integrate IT into the mainstream of education and training; To create IT awareness and ensure universal access in order to promote IT diffusion in all sectors of our national life; and
- To strengthen national identity and unity.

Specific strategies for actualizing the policy objectives included the development of a national, state and local information infrastructure backbone by "using emerging technologies such as very small aperture terminals (VSATs)[12], fibre-optic networks, high-speed gateways and broad band/multimedia technologies; provision of adequate connectivity to the global information infrastructure and the establishment of IT parks as "incubating centres for the development of software applications at national, state and local levels. NITDA, the implementing agency, was established three months following the policy announcement to "regulate, monitor, evaluate and verify progress on an on-going basis" (Federal Republic of Nigeria, 2001). NITDA was also responsible for a National Information Technology Development Trust Fund (NITDEF) to be established by the federal government with a start-up grant of US$150 million. Additional financing for NITDEF was expected to come from the allocation of 2% of the federal capital annual budget and 3% of tax on all imported finished IT products. Among other activities, the fund would be used as venture capital to provide start-up financing to small and medium scale enterprises.

Other strategies for achieving the policy's goals were the restructuring of the Nigerian education system at all levels "to respond effectively to the challenges and imagined impact of the information age" and in particular, the allocation of a special IT development fund to education at all levels; restructuring of the healthcare system by creating a national databank to provide on-line national healthcare information, administration and management at primary, secondary and tertiary levels; encouragement of "massive local and global IT skills acquisitions through training in the public and private sectors with the view to achieving a strategic medium-term milestone of at least 500,000 IT skilled personnel by the year 2004" (Federal Republic of Nigeria, 2001: v-vii). National IT awareness machinery was to be established at all levels of government while the private sector would be encouraged to participate in exposing Nigerians to the features and benefits of IT. Accordingly, the government planned to strengthen its efforts at collaborating with the private sector in the attainment of national self-reliance in the development and diffusion of IT for socio-economic goals. Finally, the policy included plans to bring "government to the doorsteps of people by creating virtual forum and facilities to strengthen accessibility to government information and facilitating interaction between the governed and government leading to transparency, accountability and the strengthening of democracy" (Ibid).

Actors

Similar to the policy on telecommunication, the NPIT consistently stresses the role of the private sector in the development of IT in the country. It defines the role of government as that of providing an enabling environment so that "ventures can flourish." It would also stimulate the private sector to become the driving force for IT creativity and enhanced productivity and competitiveness. In the chapter on government and private-sector partnerships, the policy lays out a set of strategies for achieving its goals. These include the promotion of equity participation (joint venture partnership) with IT investors both locally and internationally, and the establishment and operation of IT free zones (or IT Parks) in the country's six geo-political zones and Abuja to attract local and foreign investment in IT. Companies located in the IT free zones or Export Processing Zones (EPZ) would benefit from various incentives including reduced import tariffs, access to the Nigerian domestic market and the repatriation of at least 15 percent of their profits. Also, the government was committed to establishing "power corridors" to the IT Parks to ensure consistent and reliable power supply, as well as remove "all bureaucratic bottlenecks to the development of local capacity building" (Federal Republic of Nigeria, 2001: 24).

Table 4-1 NITDA's IT Parks

State	Cost of 100 ha land (US$) million	Cost of utilities (US$ million)	VSAT and IT services (US$ million)	Offices and Warehouses (US$ million)	Total (US$ million)
Sokoto	1.5	3.0	2.0	6.0	12.5
Yola	1.5	3.0	2.0	6.0	12.5
Jos	1.5	3.0	2.0	6.0	12.5
Lagos	4.0	3.0	4.0	9.0	20.0
Yenagoa	4.0	3.0	2.0	6.0	15.0
Enugu	1.5	3.0	2.0	6.0	12.5
Abuja	S&T Park	3.0	3.0	8.0	14.0
Total:					99.0

Source: National Policy on Information Technology (Government Copy)

Implementation of the NPIT

Five months after its creation, NITDA established an office in Abuja, the Federal Capital Territory, and began the implementation of some of its assignments. It embarked on a multi-phase public service information network with the hub at its Area 11, Abuja headquarters, with access to the internet through VSAT. Part of the first phase of the Agency's task involved setting up a computer-smart room with 30 computers in its office for the training of key government functionaries starting with President Olusegun Obasanjo and his ministers. Already, it was consulting for the Federal Civil Service Commission and some ministries in their IT development programs. Through alliances with some international organizations, NITDA helped in setting up a telecentre at the International Women's Centre in Abuja. It also tried to assist Kwale, a rural settlement outside Abuja, to link up the community with a local area network (LAN) with a future plan for an external connection to the internet through VSAT.

Part of NITDA's mandate involved the creation of IT departments in selected educational institutions. To initiate the program, the agency was expected to provide 5,700 computer systems to 185 tertiary, secondary and primary institutions within the first three years. It also planned to create training centres in Abuja and in each of the six geo-political zones in the country. Target trainees at these centres were fresh university graduates who would be taught either in the area of IT application or software/hardware development.

The objective was to re-structure the country's educational curriculum from primary to tertiary levels with a view to training half a million IT professionals by 2004.

Ajayi acknowledged two major challenges in the process of ushering Nigeria into the global information society. These were: awareness and access. Concerning awareness, he said his agency, as well as other state organizations such as the Federal Ministry of Information and National Orientation, were embarking on projects to publicize the benefits of ICTs. The message was that IT could be used to "leverage our development" as well as assist in achieving President Obasanjo's overall national goals. The agency planned IT forums in various parts of the country, and appeals to university theatre art students to incorporate the benefits of IT in their plays and story lines. A forum was also being planned for local government chairpersons and IT commissioners in the states to generate IT awareness. Creating IT awareness among the 774 local government council chairpersons was considered crucial to raising the level of IT usage by Nigerians because as Ekuwem pointed out, as soon as one local government was on the World Wide Web, others would want to score political points by also going online. In the process, the technology would spread and both usage and access would expand. Ajayi noted that the years of military dictatorship in Nigeria coincided with the revolution going on (in information technology). "We couldn't participate. We realize that Nigeria is starting late," but the country can still use IT to meet its overall objectives especially in the area of education.

Prior to the publication of the NPIT but especially since then, many government organizations had started the process of computerization. But as will be discussed in the next chapter, these efforts were still far from the goal of raising an IT-aware population. Also, policies in other sectors of the economy and society seemed disconnected from the objectives of the NPIT. For instance, while the National Policy on Education made "introductory technology" a core course at the junior secondary school level, it included "computer education" as one of five optional courses in the group of pre-vocational electives. At the senior secondary school level, students would have the option of "computer education" from a list of 18 "vocational electives" out of which they must not choose more than two. Indeed, the third edition of the national policy on education lagged behind the prevailing ethos of using "IT for education." The NPIT planned to transform the education sector at all levels—primary, secondary and tertiary—but its implementation was based on an understanding of ICTs as technologies exclusive to some sectors of the society. For instance, computers were to be made available to students in the country's

85 elite "unity" secondary schools, with no plans to extend them to the millions of other students (primary and secondary) not attending unity schools.

Contributions of the Private Sector

Through the release of the policy on ICT, the Nigerian government made the development of the technologies a priority thereby substantially increasing awareness. However, as noted earlier, initial efforts at connecting Nigeria to the global information society began outside of the government. For instance, the state did not "bring the internet" into Nigeria. Rather, it was an individual, Ibukun Odusote, who did through the assistance of some international organizations. She began reaching out to the outside world electronically in 1995. In the next few years, several groups and associations emerged. Among them were the Computer Association of Nigeria, Information Technology Association of Nigeria, Institute of Software Practitioners of Nigeria, Association of Telecommunications Operators of Nigeria and Computer Professions Registration Council of Nigeria. It was in recognition of the role of these organizations in the development and diffusion of ICTs in the country that their representatives were appointed to the initial board of NITDA. Of course, in typical Nigerian style, the sector threw up numerous associations whose activities sometimes duplicated each other thus creating room for turf wars. The myriad associations reinforced the view of Tajudeen Oyawoye, a former special assistant on IT to President Obasanjo about the deliberate efforts by some groups to mystify ICTs.

> There's a lot of money to be made from mystifying fairly simple things...The more people know about computers, the more they will know that it's no big deal...First thing to do is to get people away from thinking that this (IT) is so wonderful (whereas) it is rather mundane.

To sustain the myth about IT, a lot of Nigerians began to offer "IT solutions" with an *apriori* presumption of an existing problem. This then sparked a debate on who qualified to be called an IT professional and therefore eligible to provide "solutions" to IT users. The debate was undergirded by a move toward making the category "exclusive"—through licensing, certification and registration—even as the very nature of the technologies and the information age generally deny the exclusivity of knowledge.

State and Private Sector Alliances

Government officials in newspaper reports and in personal interviews during my research constantly made the case that "government has no business in delivering communication." It was thus divesting its interests from all communication-related enterprises. Government enterprises (such as utility companies) have been notorious for incompetence, poor services and are often viewed as drains on public resources. Thus, the belief is that if the development of ICTs is made the responsibility of the private sector, Nigerians would see wider and more intense diffusion of the technologies, and thus of their usage as tools for socio-economic development. This may indeed account for the rapid diffusion of ICTs in the country, especially in the mobile telephony sub-sector. However, given the evidence, one argues that a single sector could not have generated the growth witnessed in the ICT industry in Nigeria. The success required, and continues to do so, a strategic alliance of state and private sectors. In that sense both policy documents provided the right directions.

Also, in Nigeria personal rule and political patronage dominate—even in the current era of a democratically elected civilian government. Private (business) and public interests are therefore constantly in an alliance—often to the disadvantage of the people. Many people and companies involved in ICT are beneficiaries of government patronage through inflated contracts and services. For instance, it is common for an IT trainer to lobby for a contract to organize a "training workshop" for federal government employees, charging exorbitant fees for what often merely amount to a weekend or week away from work for employees. In another instance, Zinox Technologies, manufacturers of the first "Nigerian computer" secured the endorsement of the federal government and for sometime was its sole supplier of computers and related components. In the process, government was indirectly involved in the development of ICTs in the country often paying more than it should if participating more directly in the acquisition of the technologies and in the training.

While private-sector interests largely drive the policies on ICTs and their implementation to date, the objectives and strategies of implementation of these policies are not often integrated with other sectors of the economy. For instance, the National Policy on Education was clearly behind developments in the ICT sector. Even the most ambitious sections of the NPIT on education and human resource development overestimated the extent to which the current strategies of implementation can achieve the policy's lofty objectives and targets. Also, with so many actors on the stage, most of whose roles are contradictory and duplicative, there is need for a more coordinated effort among the different government agencies, and private sector interests as a prerequisite to successfully achieving overall socioeconomic goals.

Conclusion

In this chapter, I have discussed the two major policies on ICTs in Nigeria, their implementations and the contributions of the private sector. One notes the over dependence on the private sector for the development of the ICT sector. While this is not necessarily detrimental, there is an inherent conflict between the goals of the state and the private sector. By definition, the state (or sovereign) is responsible to the people. One of Adam Smith's duties of a sovereign is the undertaking of social works—income distribution, employment and stabilization. On the other hand, the most important objective of private sector organizations (at least those involved in the ICT sector in Nigeria) is profit maximization. Even the "professional" associations such as the Nigerian Internet Group (which claims to be a non-profit organization) are umbrella groups for people and corporations involved in for-profit ICT activities. Admittedly, these groups, such as the Information Technology Association of Nigeria, organize non-profit events, raise awareness about ICTs and donate computers to secondary schools. For sure, NITDA and especially the NCC monitor the sector closely and it is therefore not entirely a private-sector nirvana. What remains therefore is for the various government agencies to integrate the goals of the policies in this sector with their own policies in recognition of the fact that ICTs (broadly defined) are enabling technologies.

Notes

[1] Obijiofor, 2001 has an excellent analysis of the usage of ICTs by journalists in Nigeria.

[2] See Bada, 2002

[3] The paradox is that when the policies on ICTs were finally promulgated, private-sector interests pressured to make them involve as little of government as possible.

[4] Three different versions of the national telecommunications policy were released in three years – 1998, 1999 and 2000. All references to the document in this chapter come from the 2000 version.

[5] Ibid., p.23

[6] Omo-Ettu, Titi, personal interview in Lagos, December 2001, and January 2007.

[7] The third licensed operator, Nitel, did not begin its GSM operations until October 15.

[8] Personal interview, Lagos, January 2007

[9] Trans Corporation bought 75% of Nitel toward the end of 2006.

[10] Federal Republic of Nigeria, *National Policy on Information Technology*, 2001, p.iii

[11] Ajayi, Gabriel, personal interview in Abuja, November 2001

[12] In satellite communication, signals are sent from Earth to a satellite launched in space. When the signals hit the satellite, they are reflected back to earth and can be received through the appropriate technologies. According to one of my sources, Titi Omo-Ettu, an electrical/electronics engineer, satellite transmitters usually have large apertures but in recent years, it was found that very small aperture terminals could do the same job on a smaller scale but at relatively cheaper costs. In Nigeria, companies such as banks and some very wealthy individuals can now afford them. The Nigerian Communications Commission allocates the frequencies and VSAT licenses to end-users.

• CHAPTER FIVE •

All in a Day's Work: Diffusion and Usage in the Public Sector

Introduction

It sounds like one of those sermons that Jesus gave to the multitudes that gathered around him. The time is coming, and now is it, when you will walk into a hotel lobby and everyone will know who you are because you have your cell phone on you. The phone will beam signals to the hotel's central information system. By the time you arrive at the reception, the clerk already has your personal and payment information, and preferences on her computer screen. As you approach her, she says cheerfully, "Welcome, Ms. X, here's your room key, please sign here."

It is a scenario right out of the manual of proponents of a new society structured by new information and communication technologies (ICTs). These technologies will permeate every aspect of our lives, says a Nigerian civil/structural/hardware engineer and managing director of an engineering consultancy firm, Etim Amana, during a personal interview. "This is why we say ICT is the key to the future—not just the future for wealth creation, but what we do on a daily basis. This is why everyone has to join the bandwagon." That future is not quite here yet because even in ICT-rich countries such as the United States, daily life has not gone completely digitized.

However, Amana was not just fantasizing about the future: he is in a position to make it happen as a user of ICTs and an active stakeholder in the Nigerian ICT sector. He went into the ICT business as early as 1982—long before it became a "bandwagon." His firm was one of the earliest to use computers and software for engineering designs, but it stumbled into providing "IT solutions" by accident. His older brother had learnt that Shell Petroleum in Lagos was having problems with some computers it had recently acquired. He consulted with his brother and together, they were able to resolve the problem and that became the basis for a long partnership with Shell in providing "IT Solutions" to the oil company.

While actors such as Amana have been pivotal to developments in the industry, there is only so much that even they can do to integrate ICTs into all facets of Nigerian life. Indeed their "IT solutions" may help an oil company's technological and communication processes, but that is a far cry from radically transforming Nigeria's socioeconomic processes through ICTs. This was the sober realization for many ICT enthusiasts by 2008 when ICTs were yet to automatically eradicate all the problems in the country. The discourse has expectedly become more reflective. Nevertheless, in the media there continues to be a sense in which a casual observer might conclude that Nigeria is already living the ICT dream.

Certainly, there are distinct areas where growth in the sector has contributed, even if marginally, to poverty alleviation (a key objective of ICTs), as the Nigerian Communications Commission (NCC) proudly noted when it celebrated "Six Years of Telecoms Revolution" in 2007. The event marked the sixth anniversary of the award of the first set of digital mobile licenses to private telecommunications operators (PTOs). Comparing what used to be with the current landscape, the NCC noted that "the growth of the telecommunications sector is unmatched by any other sector, and it has recorded a phenomenal growth both in terms of subscribers' base and infrastructural development in the country" (NCC, 2007). While the "phenomenal growth" in the mobile telephony subsector is undisputed, the extent to which this translates to national socioeconomic development (the goal of the policies on ICT) is not precise.

This harks back to some of the core issues discussed in this book one of which concerns the process through which ICTs are expected to stimulate socioeconomic development. As indicated in the previous chapter, the Nigerian government, by formulating two major policies on the ICT sector, signalled its intentions to prioritize ICTs as tools for socioeconomic development. This raises the question: how does the government intend to achieve this objective? It is addressed in this chapter through an examination of the degree of ICT penetration and usage in selected government ministries and departments, including the office of the Nigerian president. The purpose is to understand how public officials in key ministries and departments translate the policy statements into their work practices. Data used in this chapter were collected through observation and personal interviews with several former and current policymakers and public-sector officials, many of whom were or continue to be involved in the implementation of ICT-related projects and programs in the country.[1]

The analysis of ICT usage and diffusion in Nigeria's public sector is structured by a framework developed by the Mosaic Group (Wolcott, et al, 1997).

While the Mosaic framework, appropriately called Global Diffusion of the Internet (GDIF), is used only in the study of internet diffusion in countries, it is adapted here for the examination of ICTs broadly defined. The framework consists of six dimensions that assess the development and status of internet diffusion in a country, and several determinants influencing the development and diffusion of the technology. There are two sets of dimensions. The first, which measures the extent of ICT presence and use, consists of pervasiveness, geographic dispersion, and sectoral absorption. The second set is made up of structural variables: connectivity infrastructure, organizational infrastructure and sophistication of use. Another set of variables—determinants—analyses the conditions that affect the development and current status of ICTs, and are likely to influence future developments. These conditions include government policy, culture and resources (human, financial and technology). Only three of these variables—pervasiveness, sophistication of use and sectoral absorption—are relevant to the present discussion.

Patterns of ICT Usage: Who Uses What, When and How?

Between 2000 and 2002, four government ministries and the Presidency[2] played pivotal roles in the development of policies on ICT diffusion. These were: Communications, Information and National Orientation, Science and Technology, and Education. The Ministry of Communications provided initial oversight of the implementation of the National Policy on Telecommunications (NPT) while the Ministry of Science and Technology supervised the implementation of the National Policy on Information Technology (NPIT). The Federal[3] Ministry of Information and National Orientation was chosen because of its role in disseminating information about government policies and programs. (This ministry has since merged with the Ministry of Communication to become the Ministry of Information and Communication.) The Federal Ministry of Education continues to play a fundamental role in training the human resources required to transform the Nigerian economy and position it for global competitiveness. The Nigerian Communication Commission (NCC) and the Nigerian Information Technology Development Agency (NITDA) are also included as the implementing agencies of the policies on ICTs.

The Presidency

In Aso Rock, also referred to as the Presidential Villa, Abuja (office of the Nigerian president), there was no specific IT department during the first phase of this research. Rather, many departments in different ministries claimed responsibility for getting the Presidency "connected." For instance, the Ministry of Science and Technology was in the process of building a presidential Wide Area Network to connect top government officials in Aso Rock, Ministry of Science and Technology (where the network's hub would be located), Ministries of Communication and Education and the office of the chief economic adviser to the president. However, as noted by Tajudeen Oyawoye, who served as special assistant on information technology (and later on ICTs) to President Olusegun Obasanjo, there were so many people involved that the process lacked coordination. "Everyone is offering the same information in different ways."

> Everyone wants to be in control of IT...If you ask Science and Tech, they will tell you that they were in charge of IT in the country. NITDA will tell you that they were in charge of IT in the country. (Ministry of) Information, suddenly somehow, has been able to separate the phrase information technology and take information technology to mean the technology to achieve the transfer of information and so that should be their preserve. I think my own office is the only one not really thinking that we should be in charge, although I certainly have very strong views about information technology especially in government. I have very clear ideas of what I think should be happening and where the government should be going. (Oyawoye, 2001; 2008).

When Oyawoye was appointed in May 2001, one of his primary tasks was to undertake an audit of IT usage in Aso Rock to provide a "fairly good idea of what they have there and what they were using the equipment for and the kind of people using the equipment." The audit indicated that there were about 50 computer systems in the various offices in Aso Rock.[4] Most offices had at least one computer with some having more. About 90% of the computers were used mainly for word-processing. Seven of these computers were actively connected to the internet; five were networked to each other while the other two used a different connection. Access to the internet from Aso Rock was limited to only the people who worked in the offices where the networked computers were located.

Most secretaries, personal assistants and executive officers had access to computers at Aso Rock. They word-processed speeches, letters, memos and circulars for their bosses. The more sophisticated users designed PowerPoint presentations to accompany the speeches. In some offices, such as the office of the personal assistant to the president, the computers were put to more complex functions like desktop publishing. Dial-up connection to the internet was also available in this office. In the State House Clinic, the computers were used for other things such as stock control. Some of the senior staff had laptop computers for personal use.

Table 5-1. ICTs Used in The Presidency, Aso Rock

ICTs	Used	Number	Applications
Telephone	Yes	Most offices have at least one telephone line	Conventional
Fax Machine	Yes	Unknown	Conventional
Computer	Yes	More than 100 – distributed among various offices. Ten computers were located in the "Computer Room" located just outside the Presidential Villa and used for training of Villa staff	Word processing
Printer	Yes	Most of the computers have a printer attached to them	Conventional
Photocopier	Yes	Unknown	Conventional
Internet via dial-up	Yes	Connections to the internet accessed through several networked computers.	Basic e-mailing
Internet via VSAT	Yes		

An office adjacent to the main Aso Rock building, called the Computer Room, had ten computers used mainly for training. The different departments arranged staff training courses in the Computer Room utilizing the services of external consultants. Aso Rock itself did not have standard training courses, which was one of the things Oyawoye planned to remedy.

> The idea is to see that anybody who has a computer on his desk can do certain basic things and can aspire to do more. It has nothing to do with his rank or job in the Presidency per se. I'm more interested in his competence. The person can be a secretary and be a very competent computer user, can do spreadsheets and financial scenario and things like that as long as he's good on the computer, it won't affect what he's doing. [5]

Oyawoye was dissatisfied with the minimal level of computer usage in Aso Rock. He said some of the computers were installed with Microsoft Office, but the users were not knowledgeable bout the spreadsheet or database programs that were part of Office. It was even more disappointing to find analysts who could not use scenario analysis (an application in MS Excel), and people who wrote strategic security reports and then gave to their assistants to type. To improve on the situation, he tasked himself with getting the top six people in the Nigerian government into a room for about two hours to teach them basic computer skills. The proposed students were: the President, Vice President, Chief of Staff, Secretary to the Government of the Federation, the National Security Adviser and the Head of Service. Oyawoye didn't quite achieve that. However, an arrangement with NITDA and a private consultant resulted in a 2005 agreement with Microsoft Nigeria to train 250 public officials in Abuja in a "Train the Trainer" program (Ikpe, 2005). Prior to that training, the lack of awareness of ICTs by some top government officials in Nigeria during the first two years of the 21st century was common knowledge. For instance, then Vice President Abubakar Atiku in 2001 publicly admitted ignorance of ICTs when he said at the launching of Zinox, the first "Nigerian computer" that he was not knowledgeable about IT, except that he usually saw his son *"surfacing the web all the time"* [italics added] (Amaefule, 2001). Everything had changed dramatically by the time the Obasanjo Presidency ended.

In a follow-up interview with Oyawoye in 2008, he indicated that the Computer Room had been transformed into a fully-fledged IT Unit staffed by skilled professionals. The unit was now installing, maintaining and repairing the IT assets of Aso Rock, as well as providing training for computer users in the Presidency. More importantly, all the offices now had unlimited access to broadband internet connections. Other changes included the establishment of an IT committee in the Presidency which adopted a policy of purchasing only made-in-Nigeria computer hardware except in cases where the required specifications could not be supplied by any Nigerian computer maker. According to Oyawoye:

> This policy was intended to support the then fledgling Nigerian computer hardware industry which had invested a lot of money in modern assembly lines and was finding it difficult to compete against cheap imported hardware. The policy, originally intended for the Presidency only, was approved by the President and extended to the whole of the Federal Government (Oyawoye, 2008).

The committee also proposed a policy to commission only Nigerian software/consultancy companies to work on Aso Rock computers. Said Oyawoye: "In cases where no Nigerian company had the requisite capacity to do the job only Nigerian employees of foreign companies would be allowed to have any form of access to State House/Presidency computers. This was purely a national security consideration." By the end of its mandate, the committee had also standardized Aso Rock hardware, upgrading and replacing several systems.

Ministry of Science and Technology

A key policy implementation agency, NITDA, is under the Ministry of Science and Technology. The ministry's activities have largely been driven by the idea that "information is power (and) if you don't have the right information or type, you lose out...you can't do anything without information," according to one of the scientific officers in the Ministry of Science and Technology, Abuja. In the ministry itself, there were about 50 computers with a newly acquired set of 40 that were yet to be distributed. Almost all the officers in the Technology Acquisition Department had computers (or access to one) which they used for routine assignments. The minister, deputy minister (known as the Minister of State), permanent secretary and all the directors in the ministry had computers on their desks with broadband internet access.

According to a ministry official, IT development and usage are priorities not just for the country but also in the Ministry of Science and Technology because of the importance of information technology as "a tool that can be used virtually for everything."

> It reduces the pressure on work...Computers simplify the work for us...even though we don't have enough...We were hoping that with subsequent training, we can actually get software that will make the job even more easier. What we are doing now is very elementary. We hope to improve as time goes on. IT has tremendous effects on reducing the volume of work and aiding one's schedule of duties.

Table 5-2. ICTs Used in the Ministry of Science and Technology

ICTs	Used	Number	Applications
Telephone	Yes	Unknown	Conventional
Fax Machine	Yes	Unknown	Conventional
Computer	Yes	More than 100 distributed among the top personnel and almost all the scientific officers in the Technological Acquisition Department	Word processing and task scheduling
Printer	Yes	Unknown	Conventional
Photocopier	Yes	Unknown	Conventional
Internet	Yes	Access to the internet through broadband connections	Conventional

Federal Ministry of Information and National Orientation

Among the ministries where ICT usage was observed, the Federal Ministry of Information (now part of the Ministry of Information and Communication) had the highest level of usage of many ICTs in terms of pervasiveness and sophistication of use. At the time of my research, the ministry had an IT Project

Unit headed by an officer (hereafter referred to as Mrs. O.), appointed to the position in 1999 by a previous minister of information to manage government IT projects in the country. (Mrs. O is now the director of information technology in the new Ministry of Information and Communication.) It was not long before that minister was replaced by someone who, in the officer's words, was slow in realizing the importance of ICTs. But when he did, his interest moved the ministry to a higher level of ICT usage than any government agency in the country at the time.

The IT Project Unit started off by developing a national website—an official government site to provide information about Nigeria. It also organized interconnectivity between federal information centers in the country as well as Nigeria's information centers in Nigerian embassies and high commissions. It then established a cyber café in the National Press Centre located on the premises of Radio House (which housed the then national headquarters of the Ministry of Information in Abuja). The cyber café was initially connected to the internet through a 246 KPS (kilobytes per second) VSAT. It was set up, according to Mrs. O., to grant free access to the internet to journalists who used the Press Centre. Staff of the ministry and the Federal Radio Corporation of Nigeria, Abuja (which shares Radio House with the information ministry) and members of the public also had access to the internet at the cyber café free of charge.

However, several visits to the centre showed that the queue was usually so long that it defeated the purpose of granting free access to users. Those who could afford it preferred to go to fee-based cyber cafés in the city. For most Radio House staff, it was still a positive development as they would go early in the morning, sign in, take a number, and go back later in the day to access the internet. Mrs. O. said similar cyber cafés would be established in Lagos and Jos, a Middle Belt city two hours north of Abuja. In Jos, the ministry planned to share bandwidth with the University of Jos that already had a satellite connection. And in Lagos, the ISP providing the bandwidth had a satellite connection close to the Lagos zonal office of the Ministry of Information, which would be shared with the ministry. The bandwidth sharing was expected to reduce connection costs in the two cities. Mrs. O.:

> The plan is that eventually all the information centers will be on the Internet. All of them will have computers. We've delivered (computers) to 20 of the 36 (centers) already. And the remaining 16 came in this morning and we'll do that within the next two weeks. They will be on the internet and connect to us through satellite. For now, they were doing dial up. And then the zonal headquarters will be where we'll set the info-cybers (cyber cafés) to start with.

Table 5-3. ICTs Used in the Federal Ministry of Information and National Orientation

ICTs	Used	Number	Applications
Telephone	Yes	Unknown	Conventional
Fax Machine	Yes	Unknown	Conventional
Computer	Yes	More than 150 distributed among 475 senior staff. The plan was that with the networking of these computers, all staff would have access to a computer at various points in the ministry.	Word processing, work scheduling
Printer	Yes	Many of the computers have printers attached to them	Conventional
Photocopier	Yes	Unknown	Conventional
Internet via dial-up	No	Connected to the Internet through a 246 KPS VSAT	Basic e-mailing, Web applications
Internet via VSAT	Yes		

The concept of the "information-cyber" with free access to members of the public was the first of its kind in Nigeria—"and there is none available anywhere else in the country," according to the head of the unit who said she usually walked into the centre to find saw cleaners and drivers sending e-mail to their children abroad and she would get excited about it. On access to the centre by women, she said women used it as much as men did.

> We have a register so it's possible to pick out the men and the women. I would say at that info-café, it's 50-50—we have as many women as men trying to gain access. The categories of women that we see were: mostly young girls that want to travel abroad, looking for jobs; and women who have children abroad and want to contact them by e-mail. At any point in time, you'll find that there are as many women as men otherwise we would have given some preference to women but it's not necessary because they are always there.

Within the ministry, there was also a relatively high level of ICT usage. For instance, there were about 120 computer systems distributed among 475 senior staff (Grade Level 7 to the permanent secretary). While the ratio of computer to staff was very low, it was higher than that observed in many other places. The ratio was even higher because only 75 of the staff were computer literate (with 43 being proficient and 14 both proficient and internet literate). All the secretaries, deputy directors, directors and the permanent secretary (the most senior civil servant in the ministry) had computers on their desks, or access to one. The senior staff was overwhelmingly male (60:40, in the estimates of a reliable source) because most of the women in the ministry were in the junior staff category, not included in the 475.

Mrs. O. deployed certain measures to ensure that IT usage was a constant among top officials of the ministry. One of these was forcing all directors to be IT-literate. First, she organized training sessions in which every director—starting from the permanent secretary—had to attend. And with the permanent secretary sitting through the entire sessions, the other personnel had no choice but to attend, she said. Second, when she acquired laptops, she insisted that directors who wanted the unit should turn in their desktop computers.

> There is a lot of resistance there, even though the desktops were just sitting on their desks and not being used for anything. We'd just have been adding another furniture. I've just sent a mail to them that unless they respond by e-mail they were not getting the Notebooks. I've got responses from some of them—about ten and that's really an achievement (Mrs. O).

Ministry of Communications

While the Ministry of Science and Technology oversees the implementation of the National Policy on Information Technology, the Ministry of Communications (before its merger with the Ministry of Information) supervised the implementation of the National Telecommunications Policy by the Nigerian Communication Commission (NCC) though it was often perceived that the Commission had greater powers than its supervising ministry.[6] Interviews in the ministry were focused on the departments that were most directly involved in IT policies and acquisition—the Technical Services Division and the Department of Planning, Research and Statistics. Given the importance of the Ministry of Communications in the ICT sector, the interviews were slightly different from those in the other ministries. Specifically, the questions were designed not only to assess the level of ICT usage in the ministry, but also to provide an understanding of the perceptions and expectations of ICTs by the interviewed officials.

An engineer and head of spectrum management in the Technical Services Division of the ministry (hereafter referred to as Mr. D.), explained that generally, two areas of ICT framed public discourse in Nigeria, and policies were therefore directed at them. These were computer (hardware and software) and telecommunication. And within the ministry, officials were in the process of implementing projects in these areas. A committee had earlier been set up to "ascertain the computer needs of the various departments/units within the ministry, identify their present mode of manual operation and training needs as well as recommend the most cost-effective way of accomplishing the project" (Mr. D). Other mandates of the committee included:

> Establishing the data/information processing needs of the all the departments/units of the ministry; identifying the current mode(s) of manual operations (activities) of the various departments/units in order to ascertain the operations that needed to be computerized; interacting with relevant officers of each department/unit with a view to getting information, which may be useful to the committee and the consultant; identifying the training needs of each department; recommending the most cost-effective way of accomplishing the computerization project and, identifying the present hardware and software position of the ministry.

At the end of deliberations, the committee found that the ministry had 100 computers, some of which met required specifications. Others would be replaced particularly before work commenced on the networking of all computers in the ministry, a key recommendation in the committee's report. The computers, several of which were attached to printers, had Microsoft applications with various operating systems. Several customized application packages had been installed for the ministry's management information systems. In allocating existing computers, priority was given to the various schedule officers to help them in their work, and applications installed according to the nature of their departments and units, though every computer had a Microsoft Office. The plan was to network these computers and create a local area network (LAN) accessible by all principal officers in the ministry.

According to an assistant administrative officer in the Department of Planning, Research and Statistics (DPRS), "in order to meet the modern challenges of ICTs, we were going to have many more PCs, especially in the library where all the records will be digitalized...The library will have access to the global village because we'll have internet facilities here." Eventually, the networked computers would connect to the internet and employees of the ministry would have direct access from the computers in their offices. Meanwhile, those who already had computers used them mostly as advanced typewriters. An official admitted that the computers "were very much underutilized and staff were not using them to do any other processing work as they were supposed to." For instance, the spectrum department still performed its essential tasks of frequency assignment, licensing and monitoring manually.

A major step in taking Nigeria's Ministry of Communications to the global information society, according to interviewees, was the creation of an internet domain name and website for the ministry. The website contains information about the activities of the ministry, so "our customers who may be living in Lagos, Abuja or Port Harcourt (can) access the Ministry of Communications website and know what frequencies were available and what services he or she may (be interested in) rather than telephoning or coming to Abuja to find out." When the ministry's network was fully set up and connected to the internet, the different departments were expected to compile a database of their tasks, and this information would be made available on the web.

Table 5-4 ICTs Used in the Ministry of Communications

ICTs	Used	Number	Applications
Telephone	Yes	Nine lines	Conventional
Fax Machine	Yes	Three	Conventional
Computer	Yes	At least 100	Word processing, Internet activities
Printer	Yes	More than 50	Conventional
Photocopier	Yes	At least two (with one malfunctioning)	Conventional
Internet via dial-up	Yes	Four points of access	Basic e-mailing, web applications
Internet via VSAT	Yes		

While awaiting completion of the network project, some offices in the ministry already had access to the internet. These were the Technical Services Division (TSD), the Department of Planning Research and Statistics (DPRS), and offices of the permanent secretary and minister. In the DPRS, which had three telephone lines, access to the internet was through the office of the head of the department, a director. At the time of my research, the modem was out of order and therefore the internet was inaccessible. In the TSD, with a single telephone line used for both telephone and internet connections, access to the internet was from the office of the head of the division. About eight other engineers in the division could also access the internet directly from any of the two connecting points. Since only one phone line was used, in many cases, the line was always busy and anyone wishing to dial into the internet would wait for the phone to be available. Other times, internet connections were unexpectedly interrupted when someone in another office lifted the extension to place a voice call. Each of the four internet connections in the ministry was accessible only to people working in those departments or offices (in case of the offices of the permanent secretary and minister). It could only be speculated that other staff had indirect access by receiving and sending e-mail messages through those with direct access, as observed in many offices (both public and private). It is different now as nearly all the offices have access to the internet through broadband connections. Also the ubiquity of mobile phones guarantees that staff of the ministry no long queue up for a single phone line to be available before placing calls.

Of the other ICTs, there were three fax machines in the ministry, in the TSD, DPRS and office of the permanent secretary, and at least nine telephone lines, majority of which were located in the offices of the permanent secretary and minister. Specifically about the DPRS, a staff said:

> Currently the department is not heavily equipped with ICT facilities but we are trying. Now we have about ten PCs, which were distributed among the 15 senior officers in the professional cadre (ranging from level 8 to 17). The PCs are being currently networked along with all the computers available in the entire ministry. The department is in charge of the networking because the project is under the Establishment of the Databank which is a baby of the department.

The department had one fax machine and three telephone lines for 15 officers in the senior staff cadre. These telephone lines were located in four offices (with one being a parallel line) and each of the qualified personnel who needed to place a phone call would go into one of these offices. The fax machine also shared one of the telephone lines. Access to the telephone was expected to increase when the ministry built an internal telecommunications network through a PABX system. With the addition of two external lines, it would be possible for staff without direct external connections to place and receive calls through extensions in their offices. Seven of the ten computers in the DPRS had a printer each (networked to all ten computers). The department had the only scanner in the ministry, a cable television set (located in the director's office), a photocopier that was not functional during my research, and internet connection only in the office of the director (head of the department) with a dial-up access that used one of the three lines in the department.

As noted earlier, the modem—the technology needed to connect a phone line to the internet—was defective at the time of my research and this precluded access to the internet from the office. But even when functional, access was very restricted. An officer who had access to the internet from work said he hardly used it because he had direct internet access at home, from where he did all his e-mailing—mostly writing to friends and family members living outside the country. About half of the staff in the department had regular access to the internet through cyber cafés in the city.

As in the rest of the ministry, the common usage for computers in the DPRS (with about 31 members of staff) was word processing. A staff attributed this to various factors including computer illiteracy. The ministry had prioritized the training of personnel. For some staff, the training was a good way to spend one month in Lagos at the expense of the government, at the end of which they returned to their offices in Abuja. Among some, however, there was a sense of frustration at the inability to utilize their newly acquired skills in their routine official tasks in order to increase productivity. Increased productivity was a common theme in the responses of those interviewed (both in and outside the Ministry of Communications) about the role of ICTs at the micro level. According to one official, despite the low level of usage in the department, ICTs were already enhancing the productivity of work in the DPRS by saving time, labor and energy.

> IT aids our job here in (the Ministry of) Communications. ... Several occasions, we meet to discuss face to face in meetings, but with the recent developments in computerization and networking, the time wasted moving from place to place calling meetings sitting in conference rooms, these things will be eliminated with time and this is one of the roles that ICT has played. Their importance cannot be over-emphasised. ... It is an instrument of development for other sectors.

Mr. D. added that ICTs were "very, very important" in the functioning of his Technical Services Division. He explained that the use of computers increased efficiency, reduced human involvement and therefore minimized errors, and created a faster way of transacting business.

> They enable us to communicate very quickly, to disseminate information to various users, to be able to access information (and) update (it) as things change. IT also helps us to be able to update our records periodically as the need arises, so we don't lose track of what we already have in the system and what we intend to have. It also gives us the ability to plan ahead...

Federal Ministry of Education

Interviews in the Federal Ministry of Education focused primarily on the policy environment in the context of the country's public education system. While there was no reference to ICTs in Nigeria's national policy on education, a major thrust of the policies on ICT rests on the country's ability to create an IT-literate population. According to an assistant director in the policy unit of the Federal Ministry of Education, the government had begun to integrate computer education in the curriculum of the country's unity schools and federal technical colleges (which serve as models for other secondary schools in the country).

> Computer education...is now part of the curriculum in unity schools, but yet to be reflected in the National Policy on Education...There were private schools all over the country that have computer education, and I know that a sizeable number of model

colleges in Lagos state and Kano have computer facilities...Some of the schools in Nigeria now include computer education in their curricula.

Table 5-5. ICTs Used in the Federal Ministry of Education

ICTs	Used	Number	Applications
Telephone	Yes	Two lines	Conventional
Fax Machine	?	?	?
Computer	Yes	Unknown	Word processing
Printer	Yes	Unknown	Conventional
Photocopier	?	?	?
Internet via dial-up	No		
Internet via VSAT	No		

An official in the ministry noted that computer education should ideally start from the primary school level but "the...cost of putting down such facilities (for computer education at the primary level) across the country is gigantic. It is not something that one can achieve overnight. For now, in many parts of the country, basic computer education is included in the curricula of private schools—from early childhood to secondary school education. Such schools are expensive and beyond the reach of ordinary Nigerians. The official said ICTs were too important to ignore because they have a lot of possibilities especially in distance learning and achieving the country's universal education goal.

> Everybody cannot be in the classroom. There is no way you can arrange formal education for the large population we have in a country like Nigeria. So we have to start thinking about using the new ICTs to get to the masses out there wanting to read but having no access to formal education. We've got to explore a situation where new ICTs can be used to reach a large number of people through distance learning.

While the Federal Ministry of Education had the mandate of training an IT-literate new generation of Nigerians, in its national headquarters in Abuja, there was very low usage of the technologies. On the number and kinds of ICTs used by staff of the ministry, an official simply said, "not much" and this was further confirmed by a deputy director of planning, research and statistics whose department catalogued IT acquisitions in the ministry. The actual numbers could not be determined as the assistant chief education officer in charge of IT purchasing refused to grant an interview. In July 2008, an e-mail was sent to the Federal Ministry of Education through the address on its website with a request for IT data update. A follow-up e-mail was sent a month later. There was no response to either by the time this manuscript went to the publisher in October 2008.

The ministry launched its website in 2007. This meant that it had either acquired internet access or contracted the creation and administration of the website to a for-profit organization without necessarily acquiring the technologies. This was the model for many Nigerian public organizations with websites. It was common for the "contractors" to move on to other things leaving the websites without updates and personnel to respond to requests for information additional to what was already available on the website. The Ministry of Education is therefore not the only government department in the country that does not respond to enquiries made through their websites. The Federal Ministry of Information and Communication, The Presidency, the NCC and NITDA (key departments in the ICT sector) do respond to enquiries sent to their "Contact Us" e-mail addresses either.

ICT Usage by Policy Implementing Agencies

The Nigerian Communications Commission (NCC)

The NCC had an information technology unit within months of starting business in 1994. At the time, there was only one person—who was both the founder and lone staff—of the unit. Several years and three relocations later, the unit had a full-fledged staff. Its objective was to position IT as an integral part of the Commission, and it started off with single-unit systems. Its core functions were system administration and networking, and software development. Initially, most of the Commission's staff were knowledgeable about basic typing without computer literacy, and so the IT unit drew up a training manual for everyone. In 1997, the computers in the Commission were networked and a Local Area Network was thus created. It was then necessary to re-train staff on working in a networked environment especially the security aspect of it and understanding that "being on the network does not mean that other people have access to your work," according to a senior staff of the IT Unit. The Commission moved to its permanent building in Abuja in October 2001 and re-networked its computers. Then the IT unit "introduced internet-on-demand ... and now had to educate people on how to use the internet to carry out general functions such as research." At the time of this research, there were about 120 computers (including both desktops and laptops) in the Commission distributed among 170 staff. All staff, except drivers, security people and telephone operators, utilized the computers for various purposes.

> Basically, as of today, we are working on quite a number of things, trying to build a databank for the engineers in the licensing department. A lot of information is stored in Excel and Word so we are now going into automation beyond desktop operations. We are also going to automate human resources and accounting functions. We were doing all of these in bits.[7]

A senior official was positive that the use of IT, particularly automation, had raised productivity at the NCC.

> Our core function is to regulate telecommunications, to license people to participate and create competition in the market. Let's look at the licensing or consumer affairs department. They license private telecommunications organizations but have no way of storing the information or retrieving the information when it is required by Consumer Affairs. That will create a lot of chaos and will definitely affect the productivity of the organization at the end of the day.

Table 5-6: ICTs Used in the Nigerian Communications Commission

ICTs	Used	Number	Applications
Telephone	Yes	Unknown	Conventional, Internet access
Fax Machine	Yes	Unknown	Conventional
Computer	Yes	120 distributed among 170 staff members	Various – word processing, spreadsheet, automation of duties
Printer	Yes	Most computers have a printer attached to them	Conventional
Photocopier	Yes	Unknown	Conventional
Internet via dial-up	Yes		Basic e-mailing, Web applications, research
Internet via VSAT	Yes		

Nigerian Information Technology Development Agency (NITDA)

This agency quickly established soon after it had been created in 2001. Some initial accomplishments included being connected to the internet through a VSAT—a fairly new connection infrastructure in 2001. It is useful to note that NITDA was the only government agency, besides the Federal Ministry of Information and National Orientation that was connected to the internet through a VSAT during the first phase of my research. It was also the only one that had networked all its computers. Today, its website boasts that as a "world class IT agency," it was creating a forum for people to make appointment with its officials online. Unfortunately, the "submit" button, just like many of the links on the website, was inactive, as at September 2008. Messages to the contact e-mail address posted on the website repeatedly bounced back.

Analysis of ICT Usage and Diffusion in the Public Sector

Wolcott, et al (1998) developed an analytical framework, Global Diffusion of the Internet (GDI), to assess different dimensions and determinants of technological diffusion in countries. This framework is similar, in some ways, to Kendall's stage approach (as discussed in Chapter 2). In the latter, there are five stages in the life cycle of technology—invention or discovery, emergence, acceptance, sublime and surplus (Kendall, 1999). However in the following paragraphs, the level of ICT usage and diffusion in Nigeria's public sector were analyzed within the Wolcott framework of two dimensions (pervasiveness and sophistication of use) and one determinant (sectoral absorption—the degree to which the technologies were being used by different sectors). While the Kendall framework facilitates a quantitative assessment of ICT penetration, the GDI framework enables both quantitative and qualitative assessment of diffusion and usage.

Pervasiveness

There are five levels ranging from 0 (non-existent) to 4 (pervasive) on this scale. Level 0 means that the technology is non-existent in any viable form in a country (or sector) while Level 1 (entrant) indicates the presence of the technology at a very experimental level. On Level 2, the technology has been established and is used by a small number of users, and "experience with the technology is accumulating" (Goodman, et al, 1998). The technology becomes common and pervasive on Levels 3 and 4 (respectively).

Adopting the GDI framework, the pervasiveness of ICTs in Nigeria falls between levels two and three. In the public-sector organizations studied, different types of ICTs were used at different levels of intensity and penetration. Even in the Federal Ministry of Education which indicated the least usage of ICTs, there was a general awareness of the technologies. Many of the senior staff had attended some ICT training (though concentrating mostly on basic computer literacy), and an IT acquisition unit had been created, with its officials regularly attending IT-related conferences and workshops. Support services and goods were "accumulating" through the broad range of activities undertaken by private-sector interests, particularly those in the "IT solutions" business. There was also an increasing level of local improvisation of the supporting technologies (especially software) pivotal to a higher level of ICT pervasiveness. For instance, of the more than 100 computers recorded for the Ministry of Communications, more than half were "unbranded," or clones locally assembled by Nigerians.

Sophistication of Use

This dimension highlights multiple usage of ICTs in a given country, sector or organization. There are five levels ranging from zero when the technologies are not used at all, to Level 4 when the technologies are transforming and users apply or seek new ways of increasing their capabilities. In between are Levels 1, when the "user community struggles to employ technology in conventional, mainstream applications," Level 2, when users attempt to "change established practices somewhat in response to the technology," and Level 3 when usage may result in innovations particularly as indicated in "significant changes in existing processes and practices" (Wolcott, 1997).

The sophistication of ICT use in the Nigerian public-sector organizations studied during my research can be said to be at Level 3 (innovative). While usage is still limited to most conventional mainstream applications (such as word processing) users are beginning to "change established practices somewhat in response to the technology, or in order to respond to the technology" with innovation (Ibid). Users of ICTs (such as the NCC and NITDA) have led the vanguard in sophistication of use in somewhat transformative manner. For instance, by August 2008, the NCC had created databases for its Spectrum Management and Monitory System and Legal Internet Registry Information System, both of which are accessible online by registered users. Also by 2008, NITDA had gone beyond merely networking its computers and connecting to the internet through VSAT—an innovative move in 2001 when the agency was first established—to the creation of an interactive online discussion forum on its website.

Table 5-7. Diffusion of ICTs in Selected Government Offices in Abuja

Dimension or Determinant	Level	Explanation
Pervasiveness	-3 (Established-to-Common)	This is a transition from level 2 to 3 because the different ICTs have been established in the agencies studied. In some organizations, the presence of computers (desk top and Notebooks), internet access (through broadband connections) and other aspects of ICTs and applications are common and taken for granted as regular work tools. However, the diffusion is uneven with some organizations such as NITDA being more ICT-penetrated than others (such as the Ministry of Education).
Sophistication of Use	3 (Innovative)	Some agencies used ICTs in more sophisticated ways than others but the most common applications were conventional and mainstream with some attempt at changing pre-established functions. Using the computer for secretarial and administrative duties coexisted with automation of more complex processes such as searchable online databases.
Sectoral Absorption	3 (Common)	The organizations included in this study had all the ICTs in varying degrees of sophistication and penetration. It is important to note though that these ministries and agencies belong to the same sub-sector. Their inclusion in this research was based on their involvement in ICT policy and practice in the country. The level of ICT absorption by them is therefore not indicative of the status of ICT diffusion in other sectors or industries in the country.

Sectoral Absorption

This determinant measures the level of ICT usage in the different sectors of the society. In the GDI framework, the sectors are academic, commercial, health and the government. Sectoral absorption is rated from 0 to Level 3 with 0 representing non-existent. At Level 1 (rare), less than 10% of the organizations have ICTs while at Level 2 (moderate), between 10% and 90% have ICTs, and at Level 3 (common) more than 90% of the organizations have

ICTs (Goodman, et al, 1998).[8] In rating sectoral absorption in the context of Nigeria, it must be noted that this chapter focuses just on the level of ICT usage and diffusion in the public sector, and only seven representative agencies were used for this study. However, considering the pivotal roles played by these agencies in the development and diffusion of ICTs in the country, their level of ICT usage does provide insight into what obtains in the general public sector, particularly in the federal government ministries in Abuja.

Absorption of ICTs in this sector is at Level 3 (more than 90%) if one considers ICTs in general—which include photocopiers, fax machines and telephones. Again, even the Ministry of Education, with its low usage of ICTs had at least two phone lines and some computers, the number of which could not be determined. At the other end of the spectrum was NITDA with its fully networked "computer smart" building and access to the internet through a VSAT—one of the most sophisticated connectivity infrastructures in the country. Also, one can assume that with the explosion of mobile telephony in the country and diffusion of cell phones equipped with web application protocols, absolute ICT usage has exponentially increased. While the primary purpose of acquiring cell phone lines by most civil servants is personal, in a country where the boundaries between the personal and official are not clearly defined personal phones also fulfill official functions, and vice versa.

Prospects for E-revolution

Former president, General Olusegun Obasanjo, had on numerous occasions declared that the development of new ICTs in Nigeria was a national priority. Public officials in the country, particularly at the federal level, constantly echo this, stressing the role of the technologies in giving Nigeria the leverage to participate and compete in the global economy. All the public-sector people interviewed during my research were unwavering in their belief that ICTs hold the solutions to problems of socioeconomic development in the country. They all had examples of countries and communities around the world where the development and application of ICTs had propelled them from poverty to great wealth. Frequent examples were India, Brazil and Malaysia.

The possibilities for Nigeria's e-revolution are endless. The banking sector has led the pace in actualizing the "e" concept of development in the country. Several banks have not only automated their processes but offer services that are competitive with their counterparts in other parts of the world. The technologies have worked for the sector so much that a Nigerian-based bank, Guaranty Trust Bank, was authorized by the UK Financial Services Authority in 2008 to operate full banking services in London, UK. Its advertisements are conspicuous at Heathrow and Gatwick, the two major international airports in the British city. All 25 banks in the country have websites (though many of the URLs showed broken links when I tried to access them on July 14, 2008) and offer full internet banking as well as "m-banking"—banking services through a cell phone.

A former director-general of NITDA said during an interview that though Nigeria started its e-revolution very late it had the capacity to accelerate its march toward the global information society. A major concern was the persistence of poor infrastructure in the country which continues to pose enormous hurdles.[9] But many consider this to be the strength of ICTs as technologies capable of overcoming "industrial-age" hurdles because both the mode of production and commodities have changed. Thus journalists and other IT enthusiasts spoke about how Nigeria's crafts could be sold over the internet. But this argument presupposes the ability to conduct electronic transactions, and to ship goods to any part of the world. Nigeria's postal system is one of the most rudimentary, even in the West African sub-region. While private and multinational courier companies do operate in Nigeria, they are vulnerable to myriad infrastructural constraints in the country.[10] Also, even in a post-industrial world, there are restrictions on goods of certain standards and origins from entry into the global market place. For instance, many Nigerians operating "African stores" in the United States are constantly running at a loss when relevant federal authorities refuse shipments of food products from Nigeria. The frustration for these merchants is that the rules keep shifting and they often depend on the temperament of the inspecting officer because the same products that are allowed at a given time may be rejected by another officer. A system that lacks fixed and transparent rules is also susceptible to corrupt practices by officials. Clearly therefore while the markets for information are open in a global information society, industrial-age rules continue to govern movement of goods, services and people.

An official of the Information Technology Association of Nigeria aptly described these hurdles as "industrial-age thinking" arguing that in the new digital age, traded goods will not always be physical and thus will not be vulnerable to political, economic or physical restrictions.

> The information age is proving that goods are not just physical goods. Knowledge is goods and that is why intellectual property rights come into play. Indeed at the moment, the transactions that are going on in the world at the maximal level are knowledge and software goods...It is knowledge that is being ...If science can force goods to be miniaturized, then automatically those borders that we see in the physical realm will collapse because those things we move by ships will be moved by planes.[11]

This is all possible, but until the borderless information global society become pervasive, Nigerians and their businesses are still bound by industrial-age rules that restrict movement of goods, services and labor—as many Nigerians who have tried in vain to obtain travel visas or ship goods from Nigeria to the United States can confirm.

Conclusion

In this chapter, I have presented the results of research conducted through field observation and personal interviews in selected federal government ministries and agencies in Nigeria. The research was aimed at examining how public-sector employees translate policy statements into practice through their ICT usage and diffusion. The analysis, based on the GDI framework, indicates that in the Nigerian public sector (as reflected by the ministries and agencies studied), ICT usage and diffusion have extended beyond conventional levels to common and pervasive in some dimensions. The technologies have been accepted and different agencies and personnel are innovative in their use of the different factors. At the same time, there are agencies and ministries where even at the end of 2008, the technologies are still shrouded in its novelty and are therefore hardly established. Indeed, in the second phase of the research secretaries were observed typing crime-scene reports (in triplicates, no less) on manual typewriters in one office in the state police headquarters in one of the southern states.

While Nigeria has not yet arrived at its ICT destination and all aspects of life have not gone digital (or automated) according to Mr. Amana's vision of an "e-world," the country seems far along on its journey to the global information society. This may not automatically transform the economy of the country in the manner that fuels the ICT4D discourse; nevertheless, it has enormous impact on the socioeconomic processes in the country in significant ways. For instance, even the average bank customer can expect relatively faster service than was available in the past because tellers can easily access accounts on their computer screens. Current and future policymakers may need to continue with the momentum created in the Obasanjo years. The former president moved from being an ICT sceptic to an enthusiast such that the most radical policies and developments in the sector occurred during his eight-year tenure. Interestingly, his successor, President Umaru Yar'Adua, made no reference to ICT development in his "Seven-Point Agenda" even as self-interested public officials were eager to tag their pet projects to the Agenda.[12] The ICT4D has been a powerful "pet project" for many public officials who often seem to be robotically responding to pressures from representatives of North American and European ICT manufacturing companies as well as indigenous contractors who hang around the corridors of power in Abuja. There is evidence however that these policymakers are gradually taking ownership of the process to make ICTs serve their purposes rather than those of the manufacturers. Perhaps then, Amana's vision of a fully digital life will become possible, albeit in the very far future.

Notes

[1] While all interviewees gave verbal consent for their opinions to be used in this book, only those who gave written permission are identified by their full names. For all others, either their last initial is used or their official designation at the time of the interview.

[2] The Presidency is a broad term referring to many departments and agencies directly responsible to the Office of the President. The "Presidency" is used in this chapter to refer to the offices of the president and vice president and their staff, also known as Aso Rock or Presidential Villa in Abuja.

[3] While these ministries were all "federal," the prefix is used to distinguish the national headquarters from their equivalents in the states (which are prefixed by their particular state—for example, Akwa Ibom State Ministry of Information). Some ministries, such as Communications, do not have state "branches" and often go without the prefix, "federal."

[4] A computer "system" usually consists of the computer itself (CPU, monitor and keyboard), a printer, scanner and one or two UPS (electricity back-up units). Most computer vendors, especially when their customers are governments and organizations, sell a "complete system" as a single package.

[5] Tajudeen Diekola Oyawoye, personal interview in Abuja, November 2001 and by e-mail, May 2008.

[6] Anecdotal evidence indicated that turf wars were frequent between Ernest Ndukwe of the NCC and directors in the Ministry of Communications who often felt that their positions were superior since they worked in the supervising ministry.

[7] Personal interview with a senior official of the IT Unit.

[8] The rating for sectoral absorption was developed for internet diffusion. I adapt it to the diffusion of ICTs in the public-sector departments studied during this research.

[9] These hurdles are discussed in Chapter 7.

[10] While in Nigeria for this research, I sent a package by "overnight" courier from Abuja to Calabar—a distance of ten hours by road, and given the peculiar air connections, two hours by air. It took one week to arrive. I had to personally collect it from the courier office and hand-deliver to the recipient!

[11] Personal interview in Lagos.

[12] A spokesperson for Galaxy Backbone, a federal government corporation that provides bandwidth connectivity, in a speech at a public forum linked the organization's mandate to the Seven-Point Agenda even though a close reading of the agenda does not support such linkage. Retrieved Sept. 29, 2008 from http://galaxybackbone.com.

Notes

• CHAPTER SIX •

Use IT: Patterns of Usage in the Societal Context

Introduction

There is a jaded response in the West to the presence of information and communication technologies (ICTs) in people's lives. Issues of privacy, security, identify theft, pornography and leisure time, or its absence as a result of the ever-present demands of the telephone, e-mail and other forms of ICTs, have gotten on the agenda of academic research and public dicussion. However in Nigeria as in many sub-Saharan African countries, ICTs have not yet lost their novelty. In fact, even with the ubiquity of the cell phone—with many Nigerians owning multiple hand sets and subscriptions—it may be a long time yet before Nigerians are discussing the deleterious effects and "undersides" of these technologies (Murphy, et al, 1986; Leenes and Koops, 2005). Currently in the Nigerian society, access to any form of ICTs—especially those that are as visible as the cell phone—has become a status symbol, not just of socioeconomic class but of social sophistication. The ICT vogue permeates all aspects of Nigerian urban societies, starting from the prioritization of ICTs in the national development strategies of a previous administration to application at the societal level.

The process of connecting policy with practice in the ICT for development discourse was the focus of the previous chapter. This chapter concentrates on the societal sector as reflected in the responses of college graduates' familiarity with and usage of ICTs. Essentially, the chapter presents the results of questionnaire administered to 408 new college graduates in three Nigerian cities: Port Harcourt (in the East), Lagos (in the West) and Abuja (the federal capital territory) and in two waves: August–December, 2001 and January 2007. The second wave sampled the same population but only in Lagos for reasons that will be explained later in the chapter. The questionnaire was designed to track the level of awareness, perception, usage and expectations of ICTs by a significant population in the society. It included some open-ended questions soliciting for respondents' opinions on the factors that they considered would most likely facilitate or hinder ICT development and the perceived roles of the public and private sectors. About 75 percent of the questionnaire was administered through face-to-face interviewers where the interviewer read out the questions to each respondent and recorded the answers. The other 25 percent of respondents participated through returned questionnaires.

Survey participants were selected from among members of the National Youth Service Corps (NYSC) also referred to as corps members or corpers. These are fresh college graduates under the age of 30, who, having completed at least four years of post-secondary education, are mandated by law to serve the country for 12 months in places outside their "states of origin," college or university location. The Yakubu Gowon Administration (1966-1975) started the NYSC scheme in 1973 to create national unity in a country still recovering from a bloody ethnic-centered civil war. It was primarily aimed at offering young Nigerians the opportunity to learn about the cultures of other ethnic and religious groups in the country. Under the scheme, corps members are required to report every week in the NYSC state or zonal offices for their "community development service." This provided the platform for my daily contacts with them in Abuja, Lagos and Port Harcourt during the research. Through a method of random sampling, 408 participants in the questionnaire emerged from 64 broad disciplinary backgrounds[1], representing 31 (of the 36) states of the country and Abuja. The questionnaire was successfully administered to 196 women and 184 men, with 28 not reporting their gender.

Several factors informed the selection of this group for participation in the research. First, majority of users of ICTs, especially the internet, are likely to emerge from this age group. Even in the United States, 87 percent of internet users in February and March, 2007 were 18-29, higher than among any other age group (Pew Internet & American Life Project, 2007). Secondly, I made a prior assumption that the cognitive levels of this population especially on matters related to ICTs were above those of the "average Nigerian" given that the adult literacy rate in the country was 69.1 percent (UNDP *Human Development Report*, 2007/2008). The computer literacy rate was even lower. It was therefore considered that interviewing non-college educated people might not yield much productive results. Most importantly, the NYSC group was easier to reach and controlled for religious, geographical and gender diversities.

In the second wave of the survey with Lagos participants in 2007, at least two of the rationales for choosing the NYSC population were no longer valid. The emergence of the "umbrella people" following the rapid diffusion of mobile telephony witnessed greater participation in the ICT terrain by non-college educated people. The umbrella people are "airtime resellers found in most street corners around the country, third party site engineers, roadside recharge card hawkers, handset distributors" (NCC, 2005: 37). Also, women participated at equal levels, if not greater, than men in the process of utilizing ICTs for socioeconomic growth. Many of the "umbrella people" are female, and this was mirrored by the fact that 63.5 percent of participants in the second research were female.[2] Indeed, this second batch of survey participants closely reflected the composition of the general urban society in their relationship with ICTs in many respects. They were enthusiastic about the technologies but cautious in their expectations about their utilities beyond communication and other basic functionalities.

The chapter is organized around some major themes: penetration of ICTs, awareness, usage, access, expectations and attitudes, and emerging issues. I begin with a discussion of the level of ICT penetration in Nigeria.

Penetration of ICTs in Nigeria: The Number Question

Until recently, Nigeria was not famous for its record-keeping. However, efforts by the Nigerian Communication Commission (NCC) has improved so much that there are fairly reliable ICT statistics in the country. The Commission circulates an annual survey to the various providers of telecommunication and internet services in the country. The responses become part of the industry statistics published by the Commission, as well as reported to external organizations such as the ITU. According to the NCC, there were 48,049,304 active mobile and 1,557,355 fixed phone lines in Nigeria as at May 2008. The installed capacity of both kinds of phone connections was nearly 100 million lines. The teledensity for main (fixed) lines was 1.07 percent and 27.28 percent for mobile lines (ITU, 2008). The NCC combined fixed, fixed wireless and mobile phone lines to achieve a total teledensity of 38.09 as at June 2008 (NCC, 2008).

The penetration levels of these technologies actually go deeper than the numbers might suggest as a consequence of two major factors. First, there are at least 10,000 cyber cafés in the country. According to Arnold (2005), "A cyber café (aka an internet café or PC café) is a commercial venue where members of the public can access the net for a fee, usually per hour or minute. Some cafés offer unmetered wireless access." Through cyber cafés, more people—especially urban youth—gain access to the internet and computers. Secondly, the presence of "umbrella people" who retail airtime on cell phones or fixed wireless phones on the street also facilitates access to these technologies for people who do not own phone lines and therefore are excluded from the official counts.

Table 6-1. Level of ICT Penetration in Nigeria

Technology	Nigeria	Africa
Phone lines (total)	1,557,355	30,555,000
Phone lines per 100 inhabitants	1.07	3.25
Mobile phone subscribers (total)	48,049,304	270,595,700
Mobile phone subscribers per 100 inhabitants	27.28	28.11
Internet hosts (total, 2007)	2,498	1,830,000
Internet hosts per 1000 inhabitants (2007)	0.02	2.0
Internet users (total)	10,000,000	51,982,300
Internet users per 100 inhabitants	6.75	5.43
Estimated PCs (total, 2006)	869,000	13,710,000
Estimated PCs per 100 inhabitants	0.8	1.74

Sources: International Telecommunications Union, July 2008; Livraghi, 2008.

Based on the analytical framework developed by Wolcott, et al (1997) to determine pervasiveness, a dimension of the level of internet diffusion and penetration, Nigeria is clearly on Level 3 (Common). A country reaches this level when "the ratio of internet users per capita is on the order of magnitude of at least one in a 100" (Wolcott, 1997). The framework remains valid when "internet" is substituted with "ICTs." Therefore, ICTs generally are "highly dispersed" and Nigeria is rated on Level 3 in geographic dispersion because usage has diffused to "at least 50 percent of the first-tier political sub-divisions of the country" (Ibid). The 36 state capitals constitute the first tier of political sub-divisions in Nigeria.

Awareness and Perception of ICTs

Awareness of ICTs was measured by a question requiring respondents to choose from a list of different technologies those that they considered to be information and communication technologies. These were: radio and television, telephone and fax machine, cellular phone, computer, printer and photocopier, internet—e-mail, World Wide Web (WWW)—and Other. The group of "Internet—E-mail and World Wide Web"—was routinely recognized as ICT while computers and telephones did not receive similar attention. As many as 88.2 percent of the respondents in the first phase of the research selected "Internet—E-mail, World Wide Web (WWW)," followed by "satellite and cable communication systems" 50.9 percent of the time (Figure 6-1). The third most recognized ICT was the cellular phone, at 32 percent, followed by "computer, printer and photocopier" at 26.1 percent. Interestingly, "radio and television" was selected almost as many times (22.5 percent) as "telephone and fax machine" (21.9 percent).

Figure 6-1. Awareness as Measured by Identification of ICTs

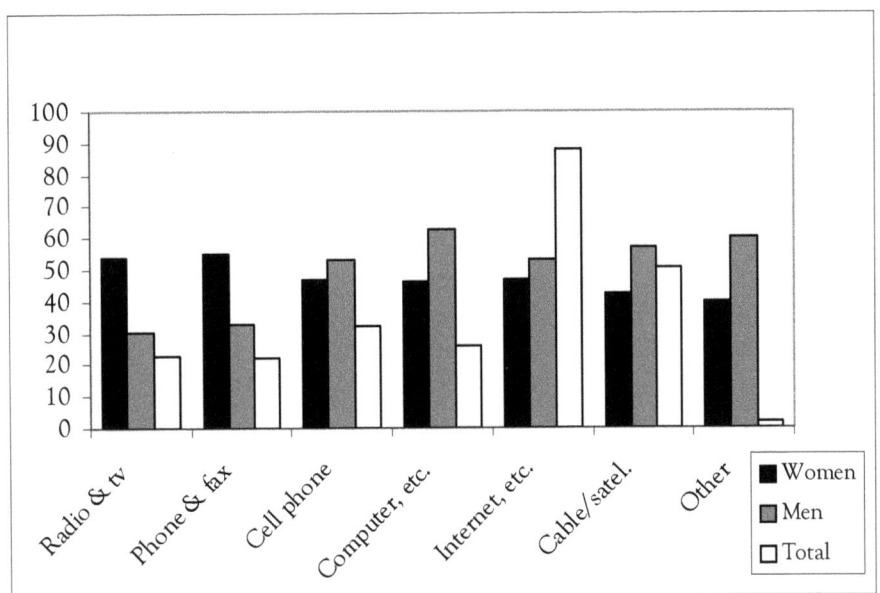

In the second phase of the survey, "radio and television" was dropped. The internet was selected as a new ICT by 77.4 percent of the respondents in the second wave of the survey while 69.9 percent chose the "GSM phone"

(which now replaced "cellular phone." Thus, more than twice as many respondents as those in 2001 identified the GSM phone as a new ICT. This was quite surprising as one would have expected that the ubiquity of the cell phone would make it so commonplace that respondents would no longer consider it new. One acknowledges that the reduction in the number of possible technologies could have constrained the choices thereby causing respondents to recognize certain technologies more times than they would have if there was a longer list of items from which to choose.

More men than women identified the core information and communication technologies (as operationalized in this book). There was no significant difference by geographical location. Besides the recognition of the "Internet–E-mail and World Wide Web" and "computer, printer and photocopying machine" more times (89.6 percent and 34.9 percent respectively, in the first wave of the research) by respondents in Lagos, the results did not indicate any geographical disparity in the exposure of youth corps members to ICTs in any of the three cities. Indeed, the margin among the cities in the recognition of the internet as an ICT was narrow. As can be seen from Figure 6-2, Lagos respondents selected the internet only 0.8 percent of the time more than those in Abuja and 2.7 percent more than Port Harcourt respondents. The only significant difference was in the recognition of "computer, printer and photocopying machine" 34.9 percent of the time by Lagos respondents as against 24.2 percent and 18.8 percent by Port Harcourt and Abuja respondents respectively. Respondents in Lagos considered the "radio and television" more times (27.4 percent) than those in the other two cities. Port Harcourt respondents seemed to stress the role of telecommunications in ICTs and chose "telephone and fax" and cellular phone more times than respondents in Abuja and Lagos. They also chose "cable and satellite communication systems" more times (55.6 percent) than respondents in the other cities. In selecting "radio and television" the least number of times (16.2 percent), Port Harcourt respondents showed more familiarity with the technologies that actually constitute ICTs, not only as conceptualized in this research, but as defined by many scholars in the field.[3]

Analyzing the data by city was informed by the prior assumption that given the uneven spread of primary technologies in the country, the city of residence would affect the degree to which Nigerians had access to ICTs. It was assumed that the larger and more cosmopolitan the city was, the greater the access to ICTs that one had. Of the three cities, Lagos is the most populous and cosmopolitan. It has more developed infrastructure partly because until 1992, it was Nigeria's capital; it remains the country's commercial capital. While there are no supporting data, it is a safe to conclude that Lagos is the richest city-state in the country even as it is also home to millions of urban poor. It was therefore expected that corps members in Lagos would have greater exposure to ICTs than those in other cities. This assumption influenced the decision to concentrate on Lagos in the second wave of the research.

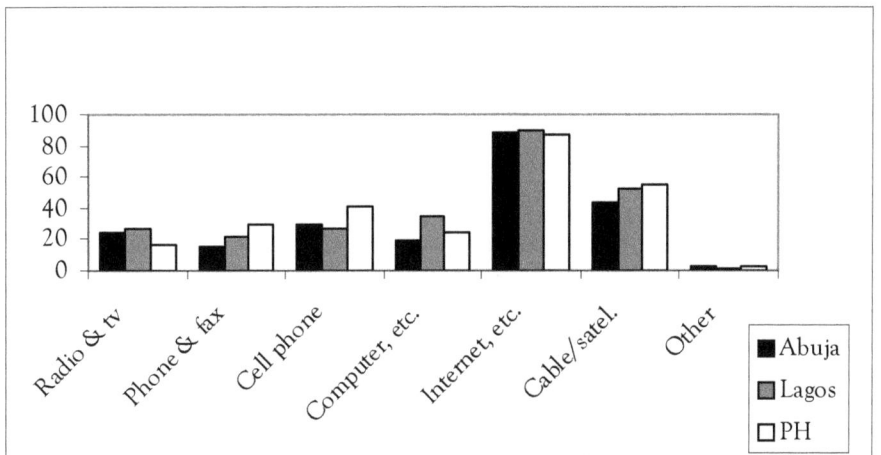

Figure 6-2. Awareness of ICTs, by City

Socially Constructed Definitions of ICTs

To evaluate the level of awareness of the technologies that constitute ICTs, respondents were allowed to choose more than one technology from the list. There was no emphasis on "new" and it was remarkable that radio and television were considered new ICTs more times by female respondents than were computers, printers and photocopiers. Many of the respondents took "information and communication" as the operative words. In this definition, it was not obvious how computer, printer and photocopier were technologies of *information* and *communication*. In processing their responses, many participants who verbalized their selections did not consider the relationship between computers and the internet.

Since the 75 percent of the questionnaires were interviewer-administered, the interviewer could observe the processes through which respondents arrived at their responses to some of the questions.[4] For instance, in deciding whether satellite and cable communication systems were ICTs, some asked: "that's CNN, right?" The interviewer would usually respond with: "What do you think?" This dialogue occurred frequently with other questions such as whether radio and television were *new* ICTs without any doubt that they were indeed information and communication technologies. Respondents said in the explanation of their choices that the telephone is not new as it had always existed. While connection to the internet through Very Small Aperture Terminals (VSATs) is a major access method in the country, most of the women who chose this as an ICT understood cable and satellite communication in terms of television broadcasting.

Through their responses participants demonstrated that ICTs lack universal definitions but are socially constructed by local contexts. This in itself was informative and allowed for conclusions about respondents' levels of ICT awareness and understanding of the issues surrounding the subject. For instance, those who earlier said they had not used the computer in the past month would later say "yes" to having used the internet, sent/received e-mail during the same period. In recognition of the presence of cyber cafés (the most common ICT access point) and the rising phenomenon of submitting handwritten letters to be sent as e-mail by cyber café staff, the question was phrased such that it did not matter if one used the computer by oneself or was assisted by another person. This was deliberately aimed at highlighting the evolving relationship between staff of cyber cafés and their clients, as observed in many of the centers. Often, there was no distinction between the person originating an e-mail and the one typing it to send at a cyber café. A client wishing to send an e-mail would sit on a chair or stool beside the staff who was either typing or in the process of sending out the e-mail. In many cases, the client would read out the hand-written letter or dictate (in the absence of written material) to the staff. As the staff was typing, the client would be reading on the screen and pointing out any errors that might have occurred. While this was going on, the two would engage in a conversation about the issues of the day—or the client might explain the background of the e-mail, and his or her relationship with the e-mail recipient. In this context, there was no distinction between the client and staff in the process of sending the e-mail.

The particular ways in which Nigerians socially constructed the meanings of ICTs were evident in other instances during the administration of the questionnaire. E-mail, particularly in the three cities of research (and among the fairly ICT savvy), was more widely recognized, and in many cases used as the generic term for the internet and WWW. This was observed through the fact that while some respondents reported that they had never used the "Internet— E-mail and WWW" as one group of technologies, they would later admit to having sent or received at least one e-mail in the month preceding the interview.[5] Many respondents in the 2001 survey expressed some uncertainties about what the internet *really* was and what online activity constituted a search. On numerous occasions, those who said they had never used the internet would later admit to sending or receiving e-mail in the month prior to their participation in the research.[6] It was also observed that those who accessed their e-mail boxes at Yahoo! or Hotmail considered the activity as "internet search."[7]

In another area, some respondents made a distinction between a "cellular phone" and "GSM phone." Again, this was a socially constructed definition generated from the intense publicity that led to and followed the introduction of the "GSM phone" into the Nigerian market in 2001. The "GSM phone" was marketed in the media as a radically unprecedented technology. Prior to this, the national carrier, Nitel, and some private telecommunications operators (PTOs) provided mobile phone services through the analogue system. Many Nigerians (including those with high levels of IT awareness) began to equate cell phones with the previous analogue technology and "GSM phone" with the new digital technology without realizing that the two were essentially the same mobile telephony but delivered through different technologies. Thus in the early days after the roll out of the digital mobile phone providers, it was common to hear people say: "You can reach me both on my cell phone and my GSM;" or "I no longer use my cellular phone. I now use GSM." Of course, with the rapid spread of the GSM cell phone in Nigeria, few would remember what the analogue cellular phone was like, especially given that fewer than 100,000 people had access to it as at August 2001 when the digital mobile phones entered the market.

Patterns of ICT Usage

This section on the patterns of usage assesses the level and nature of ICT practice among research participants. It is divided into three sub-sections: most frequently used ICTs and intensity, purposes and points of access.

Most Frequently Used ICTs and Intensity of Usage

This was measured by questions that called for the number of the different ICTs that respondents had used in the month prior to their participation in the survey, and the regularity of use during the period. More than half (55.8 percent) of respondents in the three cities had used one form of ICT or the other in the month leading up to the research (Figure 6-3). In the second wave of the survey, 75 percent reported to have used each of the ICTs on the list. Specifically, 94.4 percent of the respondents reported making at least one phone call during the period; 69.3 percent used a computer while 57.5 percent sent and 54.6 percent received e-mail. Use of an internet search engine lagged behind at 28.1 percent, followed by the percentage of respondents who had made calls using a cell phone (35.9 percent). In the second wave of the survey, 94 percent of respondents had made at least one phone call in the month prior to their participation in the research. About 71 percent had engaged in an internet activity and 54.8 percent carried out a search; 67 percent had sent e-mail while 68 percent had received. More people had used all the technologies in the 2007 research than in 2001 except for the cell phone. It was surprising that phone usage actually went down, if only slightly. At about 30.8 percent teledensity in the country, according to NCC figures, it would be reasonable to expect that all 93 youth corps members in the second phase of the research would have used a cell phone at least once in the month prior to their participation in the research.

Figure 6–3. Most Frequently Used ICTs by Gender

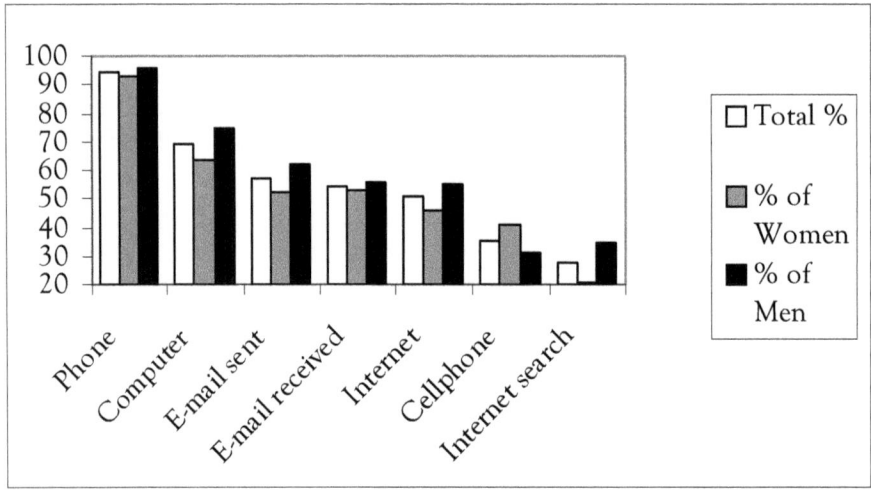

Female usage of ICTs was constantly below the total average except in cellular phone usage where they not only used it more frequently than men, but their usage was above the total average in each of the three cities of research. A significant gender gap was observed in computer usage as more men than women reported to have used it while the narrowest gender gap occurred in the percentage of respondents who had received at least one e-mail in the preceding month: 56 percent of male and 53 percent of female respondents. More male respondents reported using all the ICTs than their female colleagues, except for the cell phone, by 40 percent of female while 31 percent of male respondents made a cell phone call in the month preceding the research. This difference could be explained by the fact that in a patriarchal society such as Nigeria, people—especially men—are more willing to share their phones with women than with men. And with this knowledge, women (especially younger ones) are more likely to ask to use someone else's phone than men. Indeed, of the total number of female respondents who reported to making a cell phone call, only nine percent owned a cell phone while 73 percent used somebody else's. As cell phone usage rapidly diffused in the country, it became common for male relations—fathers, brothers, boyfriends and husbands—to purchase a cell phone for women and girls. Thus, in the second wave of the survey, and with declining costs of cell phone access, 97 percent of the women had used a GSM phone in the month prior to their participation in the research, as against 96 percent of men. Both men and women now owned their own phones and therefore frequency of usage had become a factor of ability to pay for air time. In this also, women had an advantage as it was observed that men were more likely to buy "minutes" (or recharge cards) for women than for men.

About 72 percent of women received an e-mail during the period and 68 percent sent. On the other hand, 74 percent of men received and sent e-mail during the period. Just as more women showed advantage in the frequency of cell phone usage, they also tended to access the internet at business centers (cyber café) more frequently (42 percent) than men (29 percent). Access from work or office was the second most frequent login point for women (31 percent) and men (29 percent) who used the internet in the month prior to their participation in the survey. It was not clear why women use a public internet access point more frequently than men. However, based on personal observation, I would suggest that, like the frequency of cell phone usage, regular access at a cyber café is also a function of the ability to pay. The cyber café, perhaps like a coffee shop in North America, also doubles as a social networking arena for young people. It was common to see men in several cyber cafés in Ikeja, Lagos, paying for cyber time for young women as a strategy to begin a conversation. Usually, a man would buy a cyber time "ticket" and send over to a woman through a center staff. The woman would look up at the man and smile her gratitude. After an "appropriate" interval, the man would walk over and introduce himself. Certainly, the ICT practices of NYSC members do not reflect those of the mass society. Nevertheless, there were similarities between the survey population and the larger society in various ICT practices. In this case, the same women who participated in the survey also frequented cyber cafés for their internet activities.

Cellular phone usage has increased in the country in converse proportion that of the internet. The number of cellular phone subscribers rose to 27.28 per 100 inhabitants in seven years as against 6.75 per 100 inhabitants for internet usage (See Table 6-1). There are several reasons for this including the relatively lower costs and ease of usage of the cell phone. With about ₦5,000 (approximately $45), one can obtain a cell phone, subscriber information module (SIM) and begin immediate usage on the popular pre-paid plan. Also, many service providers offer internet access through the cell phone to their high-end customers. The most rudimentary access to the internet through a cyber café costs at least ₦100 per hour and may require some level of basic and computer literacy and travel to the nearest cyber café. The cell phone, on the other hand, does not require literacy or travel time.

While the increase in cellular phone usage was expected—given the ubiquity of the technology in the country—it was rather surprising to observe that more men than women had used a computer in the month preceding the survey even though 71.1 percent of female respondents reported to being computer literate, as against 67 percent of male respondents. An equal number (78 percent) of male and female respondents said they had computer training. The assumption was that more women would report to using the computers because as in many other countries, women are predominant in the "pink collar" clerical and administrative positions. Indeed in many offices visited during my research, it was more common to find women than men behind the computers. Also there was a significant presence of women in the business centers visited. While female corpers do not represent the larger population of women in the country, it was assumed that their pattern of ICT usage would closely mirror that of the larger society.

Geographically, the highest level of ICT usage among respondents occurred in Lagos where 59.6 percent of the respondents had utilized one form of ICT or the other in the preceding month. This was followed by Abuja (56.6 percent) and Port Harcourt (46.9 percent). All 101 respondents in Abuja had made a telephone call at least once in the month prior to their participation in the survey with 96.2 percent in Lagos and 86.9 percent in Port Harcourt. Respondents in Lagos recorded the highest usage of all the technologies but they trailed Abuja and Port Harcourt in the number of respondents who had received at least one e-mail during the period. About half of Abuja respondents (50.5 percent) and 49.5 percent of Port Harcourt respondents had received e-mail during the period, as against 35.8 percent of Lagos respondents. Interestingly, at 64.2 percent, more of the Lagos respondents had sent e-mail than those in the other cities (59.4 percent and 48.5 percent for Abuja and Port Harcourt, respectively). Lagos also took the lead in cellular phone usage during the period with 44.9 percent of the respondents having made a cell-phone call in the month preceding the survey. This was followed by Abuja at 40.6 percent and Port Harcourt at 21.2 percent. In all, 35.6 percent of all 408 respondents who participated in the questionnaire had used a cellular phone (to make or receive calls) during the one month prior to the research. The least used ICT was internet search at 27.8 percent (35.5 percent in Lagos, 26.7 percent in Abuja and 21.2 percent in Port Harcourt).

Many of the criticisms about the ICT4D discourse are directed to its insistence that the mere acquisition of the technologies would usher in socioeconomic development that benefits all participants equally as in a Networld scenario (Howkins and Valantin, 1997). However the geographical disparity in ICT usage in this survey indicates that the spread of the benefits of ICT application will be uneven and access to the technologies will be determined by one's geographical position in the society. As noted earlier in this chapter, Lagos, a city-state of 13 million people is the most infrastructurally advanced in the country. Even its suburbia is more developed than many state capitals in other parts of the country. As the former national capital and home to Nigeria's formidable international sea and airports, it is the hub of socioeconomic life in the country. While not as populous, Abuja is the current capital city of Nigeria. Many youth corps members lobby and bribe to be posted to Lagos or Abuja because of the opportunities for employment and business that exist in the two cities. The higher level of ICT usage in each of the cities in the survey therefore demonstrates the significance of each of these cities in the political economy of Nigeria. The challenge therefore for policymakers is to formulate strategies to expand ICT usage outside the core centers in the country.

Purposes of ICT Usage by Respondents

The policies on ICTs in Nigeria raised expectations about the capacity of ICTs—in their usage and diffusion—to stimulate socioeconomic growth in the country. To measure how usage reflects stated intentions of these policies, the questionnaire included questions about respondents' reasons for using the various ICTs. Specifically, respondents were asked to state the purpose the last time they used each of the ICTs. The options were "personal," "official/work/business related" and "both." An overwhelming majority of respondents used the technologies mainly for personal purposes. However given the Nigerian environment, many could not distinguish between personal and official reasons. For instance, many understood "business" to apply to a privately owned enterprise such as a contracting or trading business, and "official" to refer to anything relating to their offices or places of work. Hence, respondents who used the computer to type their résumés or job applications would choose "personal" as the purpose for which they had used the technology.[8]

Table 6-2. Purpose of ICT Usage

City of research	Phone		Cell phone		Internet		E-mail sent		E-mail received		Internet search	
	Pers.	Offi.	Pers.	Offi.	Pers.	Offi.	Pers.	Offi.	Pers.	Offi.	Pers.	Offi.
Abuja	88	13	33	8	40	3	32	28	28	23	22	5
Lagos	80	24	37	12	44	2	66	4	57	2	25	5
Port Harcourt	74	12	17	4	30		46	1	41	1	12	3
Total in #	242	49	87	24	114	5	144	33	126	26	59	13
Total in percent	83.7	17.0	78.4	21.6	73.1	3.2	82.3	18.9	75.4	15.6	67.0	14.8

Pers. = personal; Offi. = official
Percentages may not equal 100 because of rounding.

As Table 6-2 shows, 83.7 percent of the last phone calls made before participation in the survey were reported as personal, as were 78.4 percent of the last calls made on a cell phone. This number increased to 87.4 percent in the second wave of the survey. Similarly, 82.3 percent of the last e-mails sent and 75.4 percent of the last e-mails received were said to be personal. The balance was distributed between those who used the technologies for both personal and official reasons and those for whom "I don't know" was recorded.[9] This was typical of the e-mail activities of respondents because the pattern of distribution between personal and official purposes was the same for e-mails received or sent in any given month.

A major thrust of the policies on ICTs in Nigeria concerns the need to use the technologies to generate economic growth internally, as well as give the country an advantage in the global economy. To understand the connections between this national goal and respondents' use of ICTs, questions were also posed concerning the destinations and origins of last e-mail messages and phone calls. As Table 6-3 indicates, most of the last phone calls (either from fixed or mobile lines) were to numbers within the country, with local and state calls being in the majority, followed by national calls. Only nine percent of respondents reported that their last telephone calls were to numbers outside the country while the last cellular calls of three percent of the respondents were to international numbers. In the second wave of the survey, 65.9 percent of the last phone calls made from a mobile phone were to local numbers. This trend may not be a factor of lack of interest in external connections as much as it is that of the cost of making international calls. In many cases, a cellular phone call to an international number costs as much as ₦90, approximately 85 US cents per minute though some of the fixed wireless telephone companies have been able to offer lower off-peak rates. Still, even with this relatively low cost, and assuming that a lot of people have access to a fixed wireless line, they would need to have personal or business relationships with people outside the country to require an international call. This is why external-oriented communication was higher with the internet because the platform does not necessarily require a two-way process.

Table 6–3. Destination and Origin of Last ICT Activity

City	Phone call			Cell call			E-mail sent			E-mail received			Type of information searched for on the internet	
	State	Nat'l	Int'l	State	Nat'l	Int'l	State	Nat'l	Int'l	State	Nat'l	Int'l	Local	Int'l
Abuja	45	53	2	16	24	1	7	23	28	5	21	25	...	23
Lagos	76	16	9	105	31	14	50	26	59	51	31	57	7	29
Port Harcourt	34	50	1	11	9	1	6	23	18	7	24	17	3	20
Total [1]	155	119	12	132	64	16	63	72	105	63	76	99	10	72
percent [2]	54.1	41.6	4.2	62.2	30.1	7.5	26.2	30	43.7	26.4	31.9	41.6	11.4	81.8

[1] Percentages are based on the total number of respondents who had used the particular ICT in the month just before their participation in the research.
[2] Percentages may not equal 100 because of rounding.

About 43.7 percent of the last e-mail messages sent were to addresses outside the country, while 41.6 percent of the last e-mail received came from addresses outside the country, followed by 58.4 percent for e-mail received from addresses (or senders) within the country.[10] These numbers were lower in the second phase of the survey. While most e-mail were sent to international addresses, the percentage was nearly evenly distributed with those sent to local addresses. Also fewer e-mail messages (23.7 percent) were received from outside the country. Of those who had done an internet search in the month prior to their participation in the research, their last searches were for international information, with information defined as "international" if it related to activities or interests outside the country. For instance, many of the respondents were seeking information on employment and graduate studies in Canada and the United States. In the section that called for the nature of the information sought (or keywords), many respondents said the following:

> Career opportunities at a company website; Graduate studies; countries in the past, Europe, France, History, French Revolution; career, sponsorship, latest news at CNN.com; DV-2007 American Visa; Graduate schools in the US; GRE requirements for admission, graduate GPA requirement for the different good schools; employment offer in Germany; online shops in the US; opportunities for graduate education abroad; immigration process to Canada.

There were also a few—11.4 percent—respondents whose last internet searches were for local information, with information considered local if it directly affected the respondents in their places of residence. Thus a respondent's search for information on how to bake was classified as local. So was the search of one respondent for information on airplanes as his company was planning to acquire one. (He said he worked in the acquisition department of an airline.) Other searches defined as local included:

> information about my project; information on GSM handsets; software drivers and downloads; information on artists and music; programming codes, congratulatory cards, electronic greeting cards; information concerning health and entertainment; information on dieting and fashion.

Points of Access

A major issue that emerges in the discourse on ICTs as tools for socioeconomic development in countries such as Nigeria has been access—as will be discussed later in this chapter. It was therefore necessary to know the level of access to ICTs that survey respondents had. They were asked questions concerning the places where they access particular ICTs—and the cost of such access. Specifically, the question was: "The last time you used (or accessed)... (name of technology) where was it?" The options were home, work (office), school, public pay phone or business center. Access to the telephone and computer was mostly from work (Table 6-4). In 2001, the second most common point of access was the home, followed by a payphone. And among those who said they had accessed these technologies from home, many said they did not directly own the access. Often, they went to friends' or relations' homes to use the technologies—especially the telephone. Others lived in homes where another member of the household owned the technology (or technologies).

With just 17 percent of those who had used the computer in the month preceding the research owning their computers or living in homes where they were available, access to the computer was invariably more at work and business centers than anywhere else. The cyber café was also the point of access for 64.3 percent of respondents who had been on the internet in the month leading up to their participation in the survey. This was followed by 28.7 percent of login from work and 6 percent from home. While 70.3 percent of all respondents in the research had their own e-mail addresses, only 24 percent respondents had direct internet access from home, and 76 percent from work. This pattern remained unchanged in the second wave of the survey. The most common point of access remained the cyber café (at 34.4 percent) while 29.9 percent used the internet at work. Only 14 percent said they had access to the internet at home.

Table 6-4: Points of Last ICT Access

City	Telephone			Computer			Own computer	Internet log in			Internet access		E-mail address
	Home	Work	Pay phone	Home	Work	Cyber café	Yes	Home	Work	Cyber café	Home	Work	Yes
Abuja	32	47	19	11	44	13	12	4	21	27	3	27	67
Lagos	36	47	11	13	51	19	19	4	20	39	10	28	86
PH	12	34	32	7	33	16	5	1	4	35	6	6	62
Total [1]	80	128	62	31	128	48	36	9	45	101	19	61	215
percent [3]	30	47	23	15	62	23	17 [2]	6	29	64	24	76	70 [2]

[1] All calculations, except the entries on e-mail address and computer ownership are based on the total number of respondents who had used the particular ICT in the month just before their participation in the research.
[2] Calculated from the total number of participants in the research.
[3] Percentages may not equal 100 because of rounding.

Given the nature of the Nigerian society—namely low penetration of ICTs and communal ownership of properties—the question about access was phrased such that there was no emphasis on "ownership." Here again, "universally" accepted definition of access was locally reconstructed to capture the social context. In Nigeria, access to certain properties (communally or individually owned) is frequently equated with ownership, thus many respondents who said they had direct access to the internet actually meant access to accounts owned by or accessible to friends and relations. Access to the internet at business centers or cyber cafés was also reported as ownership because the patron of these centers paid for time to "browse" and had monopoly of the use for the period (and therefore could claim ownership) of the technology.

As Table 6-4 shows, ICT usage by respondents occurred mostly at work and this point is significant for two reasons. First, 76.7 percent of all ICT activities undertaken were personal. And in 44.6 percent of the time, these activities were undertaken at work. Second, as has been seen, the most frequently used ICT is the telephone, followed by e-mailing activities—sending and receiving. The evidence is not conclusive; however, one might tentatively suggest that for the respondents—and many others encountered in the course of the research—these technologies are important only as they enable information gathering and communication geared toward building personal relationships and improvement in personal conditions. It is also significant that much official work time is spent on personal communication. Indeed, I found many instances of this during the two phases of the research. While I was being attended to by a teller at a bank in Uyo, a south-eastern state capital, a cell phone rang. It was the teller's. He stopped processing my transaction to take the call. Identifying who it was, he pushed back his chair and eased into a relaxed position. From his side of the conversation, I gathered he was speaking with a woman and that the relationship was "close." After about six minutes, the teller blew some kisses into the phone and then returned to me. Clearly, there were no restrictions to personal communication by on-duty bank officials even as customers were asked by security men to switch off their cell phones before entering the bank.

While the connections do exist, at the moment, the current motivations for the utilization of ICTs seem removed from the goals of national socioeconomic growth and increased productivity at the micro level. Also, actual usage of ICTs by respondents seems disconnected from their expectations (discussed in the next section) about the potentials of ICTs as tools for socioeconomic development. Respondents expect much from ICTs, but they do not equate their ICT practices with these expectations. Though the respondents in both phases of the survey did not make such connections, it is possible that personal use is a prelude to better familiarity with the technologies and, ultimately economic use.

It might be argued that given the socioeconomic level of respondents and their marginal participation in industry or policymaking, their more personal usage of ICTs is predictable and not in any way suggestive of the trend among the general population, particularly those in the vanguard of the discourse on ICT for development. Also, youth corpers are a little above the level of sophisticated interns who ordinarily would not be given access to organizations' core ICT applications. But this tendency was also observed in the interviews with public officials. While they expressed optimism about the capacity of ICTs to increase their individual work productivities and eventually lead to macro socioeconomic growth, this was not reflected in their ICT practice. Only few of the policymakers interviewed used the technologies available to them for official purposes. For instance, one high-ranking official said he used his official e-mail access to communicate with friends and relations living outside the country.

Cost of Access

Cost of access to the technologies should be a major determinant of their diffusion in a country such as Nigeria with high levels of poverty. This was a common theme in the comments of respondents in the open-ended section of the questionnaire. While there was unanimity concerning the high cost of access, the different constructions of access mediated any clear determination of how much respondents spent and on what level of access. Those who interpreted "direct personal access" to the internet as their ability to pay to use the technologies at business centers and cyber cafés spoke of access in terms of the per-minute-rate they were charged. And with this understanding, they reported that it cost them approximately ₦50 to ₦300 each time they accessed the internet—depending on how long they were online and the per-minute rates at the particular cyber café. In Lagos, the average rate was ₦120 for one hour in 2007. As at 2008, the cost of access to the phone had also dropped as a consequence of the entry of many providers into the digital mobile market through the unified licensing regime that came into effect in 2007. Mobile phone users could therefore access international numbers for as low as ₦20 per minute. Ironically, the cost of local calls remained high.

Sophistication of Use

Sophistication of use is one of the dimensions in the GDI framework adapted here for the analysis of the diffusion of ICTs in Nigeria. It highlights the number of people who use ICTs, points of access and purpose.

> Of particular interest is the "elbow" reached when the service is mature enough to attract interest and use outside the narrow community of technicians. A second major milestone is reached when the user community transitions from only using the (technology) to creating new applications, sometimes eventually having an impact on (use of the technology) elsewhere (Goodman, et al, 1998).

The levels of measurement for this dimension range from 0 (non-existent) to four (transforming). In between are levels one (assisted), two (conventional) and three (innovating). So far in this chapter, I have presented the various ways, places and purposes of usage of the different ICTs by participants in the questionnaire portion of the research. But the questionnaire also specifically inquired concerning the particular activities engaged in the last time an ICT was used by respondents. For the computer, the question was posed: "The last time you used a computer either by yourself or through some other person, what did you do?" The options were: typed a personal letter, typed an official letter, typed an e-mail, typed a résumé (or CV), typed a document related to your work or business, prepared a school project (term paper or thesis), used it to carry out your routine office work, played a computer game, accessed the internet, other. Respondents were allowed choice of more than one activity. A similar question was asked concerning the last time the respondent used the internet during the one month prior to participation in the research. The options were: E-mail (receiving and sending), internet search, browsing and Other.

Typing an e-mail message was the most common usage of the computer, by 53.3 percent of respondents in the month prior to the research, followed by typing of personal letters by 42 percent of the respondents. The third most common use for the computer was to access the internet, by 37.7 percent of the respondents. About 34 percent of respondents used the computer to type official letters; 30.7 percent did routine work while the same number played computer games. Three percent of respondents engaged in activities outside the list, but mainly for learning purposes—typing, how to use the internet, how to design a web page and trying out new software packages. Of the respondents who had used the internet in the month preceding their participation in the survey, 63.6 percent reported that their last internet activity was e-mailing (receiving and sending). The pattern was the same when analyzed by gender and by city of research: e-mailing was the most common usage of the computer and internet. In the second wave of the survey 85 percent of respondents had sent e-mail and 81 percent had received e-mail in the month prior to their participation in the survey. About 73 percent had performed an internet search. Of the 87 percent of respondents who used their GSM phone, voice communication was the most frequent activity (58.1 percent) followed by text messaging (35.5 percent)—the later being a category that was absent in the first phase of the survey. As at 2001, cellular phone usage was basic voice communication without much sophistication. This changed following the rapid diffusion of the technology creating room for more sophisticated applications. Text messaging, photographing, e-mailing, games and mobile banking competed with voice communication on the cell phone.

The results show that respondents' usage of many ICTs was conventional—Level 2 on the GDI framework. While more than half of them had used one form of ICT or the other, established practices were not altered or challenged. The overwhelming engagement in e-mailing only replaced traditional letter writing as means of communication—except now the process was faster and relatively cheaper. Thus the technologies, as currently applied by the respondents, are not innovating. Users do not push the boundaries by exploiting more sophisticated ways of using the technologies. They are also seemingly reluctant or unable to use the technologies toward the achievement of the national goals on ICT policies. Paradoxically, the lack of innovation in the use of the technologies did not rein in the expectations that respondents have concerning the capacities of ICTs.

Attitudes and Expectations About ICTs

At the beginning of the questionnaire, respondents were asked to choose three issues from a list of ten considered to be the most important socioeconomic concerns in the country. The issues were: unemployment, inflation, low literacy level, corruption, poor health services, and inadequate infrastructure, health services, and inadequate infrastructure—such as electricity, communication facilities and pipe-borne water, bad roads—hunger, crime and bad leadership. Respondents—across city and gender—consistently chose unemployment, corruption and bad leadership as the three most important socioeconomic concerns in the country (Figure 6-4). The same issues also surfaced in the second wave of the survey indicating that not much had changed in the six years since the first part of the research was conducted. Three other issues—inadequate infrastructures, crime and low literacy level—were also selected. In Abuja, 15 percent of female respondents chose inflation as one of the three most important socioeconomic concerns in the country. Generally, unemployment was chosen 80.1 percent of the time in all three cities, while corruption and bad leadership were selected 59.2 percent and 41.8 percent times, respectively. In the second wave of the survey, unemployment was selected 84 percent of the time (and by 74.2 percent of the female respondents) as the top most important socioeconomic concern in the country. This is higher than the figures from six years earlier.

Figure 6–4. Most Important Socioeconomic Concerns (Percentage of Times Selected)

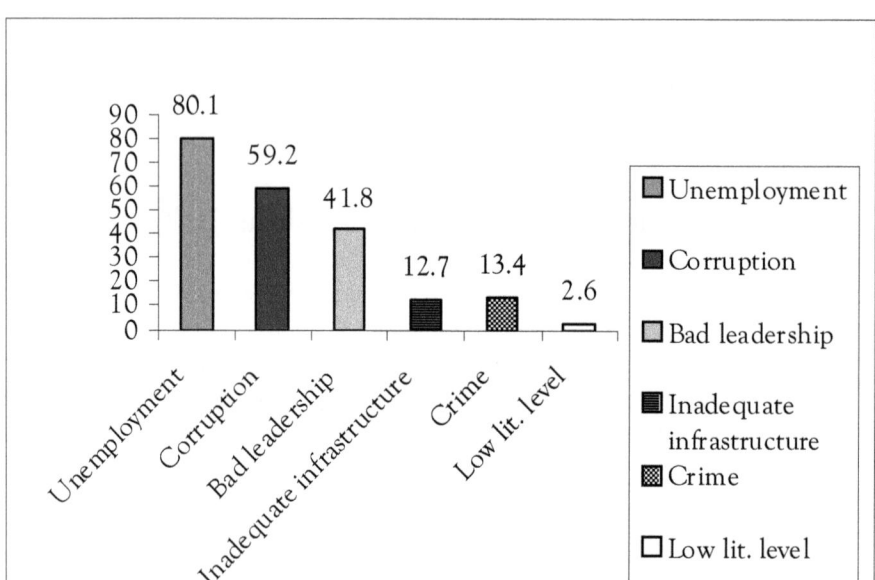

The ten issues were deliberately selected to reflect the expressed objectives and aspirations of policymakers and Nigerians concerning the capacities of ICTs to generate socio-economic development in the country. The issues were used in constructing the statements in the matrix section of the questionnaire, aimed at understanding the ways in which the statements of policymakers, especially as contained in the policies on ICTs, connect with respondents' personal beliefs and expectations about the capabilities of ICTs. This section was also aimed at providing the connections between the most important socioeconomic concerns in the country and the role of ICTs in effecting changes. Respondents were required to indicate if they strongly agreed, agreed, neither agreed nor disagreed, disagreed or strongly disagreed with each of 11 statements beginning with "ICTs will…"

Table 6-5a. Expectations About ICTs (Percentage* of Respondents, 2001)

ICTs will ...	SA	A	NA	D	SD
Stimulate socio-economic growth	52	35	8	3	0
Create employment	38	41	13	5	1
Facilitate health delivery	36	44	12	5	1
Raise literacy rates	32	31	15	13	5
Check corruption	9	18	34	25	10
Check inflation	10	24	32	21	9
Reduce crime rate	14	28	19	24	12
Eliminate hunger	8	13	28	35	14
Solve the problem of bad leadership	12	25	18	28	13
Politically empower Nigerians	20	33	21	18	5
Improve my standard of living	42	40	9	5	1

Legend: SA – Strongly Agree; A – Agree, NA – Neither Agree nor Disagree,
D – Disagree, SD – Strongly Disagree
* Percentages may not equal 100 because of rounding.

More than half of the respondents strongly agreed that the new ICTs would stimulate socioeconomic growth in Nigeria, while 35 percent simply agreed (Table 6-5a). Respondents also strongly agreed (by 38 percent) and agreed (by 41 percent) that new ICTS would create employment, while 42 percent strongly agreed ICTs would improve their standard of living. About 32 percent of respondents strongly agreed (and 31 percent agreed) that the technologies would raise literacy rates in the country. Many were however ambivalent about the capacity of the technologies to check corruption and inflation, two other issues respondents identified as important socioeconomic concerns in Nigeria. Many respondents wondered how ICTs could check official corruption and looting of public funds by government functionaries, with some suggesting that the technologies might actually create new avenues for corruption in the country and other crimes. This has already happened as scam artists (commonly called "Yahoo boys" in Nigeria) from many African countries have found the internet as a veritable platform for their activities. This is discussed further in Chapter 8.

There was much optimism about the capacities of ICTs for various aspects of socioeconomic life in Nigeria in 2001. This was not misplaced given the priority that the Obasanjo Administration conferred on the sector. Officials of the Administration, such as the president himself, began in 1999 as unwilling ICT advocates but ended up being super champions of the technologies as agents of socioeconomic development. In the first two years of the administration two pivotal policies were formulated to give official oversight to the sector. Also, digital mobile licences were issued to four operators to commence the provision of mobile telephony. Within the first three months of the commencement of the cellular phone services, Nigeria was well on its way to finally solving the country's perennial telecommunications problem.

It was in this optimistic environment and period when so much appeared possible that the first wave of this survey was conducted. It was therefore not surprising that respondents expressed much faith in the promises of ICTs to stimulate socioeconomic growth, create employment, facilitate health delivery, raise literacy rates, improve respondents' standard of living and empower them politically. While they were so positive about the capacities of the technologies in various aspects of life, they were critical enough (even then) to express cynicism about the capacities of ICTs in being the panacea to every socioeconomic and political ailment in the country. For instance, the percentage of respondents who believed that ICTs would check corruption, eliminate hunger, check inflation, reduce crime and solve the problem of bad leadership were low. In this, it was worth noting that respondents did not make any connection between reduction in inflation, corruption and crime with socioeconomic growth; or how hunger might be eliminated if the economy was vibrant and employment was created through utilization of ICTs. It is possible that respondents were echoing many of the assumptions about ICTs in the media especially as embedded in the ICT and telecommunications policy documents. The ICT policy, for instance, has a long section prefaced with "Use IT" with the "IT" serving as a pun on "information technology" and the "it" pronoun. The policy was released in 2001 with widely publicized imperative for Nigerians to "Use IT" as tools for solving many problems confronting the country. Thus, ICTs were still wrapped in myths in 2001 creating a disconnection between the realities of the technologies and respondents' expectations of them.

This assumption is undergirded by the responses of the same target population six years later. As Table 6-1 shows, the level of ICT penetration in Nigeria deepened exponentially during this period. For instance, cellular phone lines increased from 300,000 at the end of 2001 to more than 48 million in 2007. Many Nigerians and the respondents were therefore more familiar with the technologies in 2007 than they were six years earlier. Perhaps, this familiarity generated more critical attitude toward ICTs such that expectations about their capacities were more sober. It is also possible that as new college graduates, respondents were more focused on the increasing inability of the system to absorb young school leaders into the labor force. Indeed, many are beginning to point to this difficulty as explanation for the increasing incidence of crime (such as kidnappings) and proliferation of youth-organized crime in the country. As Table 6-5b indicates, 12 percent fewer respondents strongly agreed that ICTs would stimulate socioeconomic growth while eight percent less agreed. About 32 percent (against 38 percent in 2001) strongly agreed that the technologies would create employment and 33 percent (as against 41 percent in 2001) agreed. A more significant difference occurred in the number of respondents who believed that ICTs would improve their standard of living. In 2001, 42 percent strongly agreed and 40 percent agreed; in 2007, 18 percent strongly agreed and 29 percent agreed.

Table 6-5b. Expectations About ICTs (Percentage* of Respondents, 2007)

ICTs will ...	Strongly Agree	Agree
Stimulate socio-economic growth	40	27
Create employment	32	33
Facilitate health delivery	13	27
Raise literacy rates	29	26
Check corruption	9	16
Check inflation	6	11
Reduce crime rate	10	17
Eliminate hunger	6	17
Solve the problem of bad leadership	6	17
Politically empower Nigerians	8	16
Improve my standard of living	18	29

*Percentages may not equal 100 because of rounding.

Common Themes

The open-ended section of the questionnaire provided the opportunity for respondents to expand on their ICT usage patterns with emphasis on the factors that are likely to increase their level of ICT practice. Specifically, respondents were asked: "In your opinion, what factors or conditions will promote the rapid diffusion and use of new ICTs in Nigeria?" The responses are presented in the next section. For now one discusses the three major themes that emerged from the open-ended section of the question concerning respondents' aspirations about ICTs.

Employment as a major source of anxiety (and therefore the object of much optimism) was expected. As already explained, youth corps members are recent college graduates who do a one-year national service after graduation. For most of them, the year of national service is the only guaranteed employment (and therefore source of income) in the foreseeable future given the high employment rate in the country. Unemployment is therefore a major concern for the respondents. This explains why 80.1 percent of respondents (and 89.7 percent in the second wave of the survey) said it was the most important socio-economic concern in Nigeria. They expressed the hope that ICTs would create employment for them at a very personal and basic level. There were responses such as:

- New ICTs will affect my life in terms of creating job opportunities.
- It will broaden my knowledge (through) getting to know more about other professions thus giving me an advantage of getting a better offer in terms of employment.
- Through ICTs, I can get employment instead of sitting at home.
- There's development everywhere. As a civil engineer, I'll know about job opportunities.
- I hope to acquire my system very soon. It will stimulate my economy. It will create job and employment for me. It will make me to be creative. Generally my standard of living will improve. It (will) give me awareness and (and make me) highly educative.
- I believe that from them (ICTs) I can get enlightenment and improve on my knowledge and make myself marketable.
- With ICTs being the current rave of the moment, a lot of opportunities, careers and employment are just within my reach.

Information was the second theme around which participants' expectations about ICTs were framed. Generally, respondents expected that ICTs would keep them informed about events at home and abroad.

- Positively, ICTs will make me informed about events at home and abroad. (They will) improve my skills.
- (ICTs) will make communication easier ... Information is gotten easily and fresh (news) – that is on cable and Internet.
- The ICTs will make the world a global village because there will be easy access to new information.
- (ICTS) will increase communication efficiency. Access to information; possibly better job opportunities.
- The new ICTs will personally affect my life in the next five years due to the fact that information required will be easily reached in terms of (movement, standard of living, employment, etc.)
- It will enable me search and receive information from other jurisdiction that will help me increase in knowledge in my field of study.
- I personally believe that ICTs will affect me positively in the next five years because a better-informed person is better guided.

- The world is now a global village due to ICTs so it is going to affect my life in all aspects. I can learn about everything I want to and visit everywhere I want to without leaving my locality.
- If I have the right knowledge as well as access to the new ICTs, it will get me exposed to so many ideas and practical ways of being a success and contributor to the society I find myself when I apply to my course of study – Animal Science.
- Information makes the world go round. Personally new ICTs will enable me get informed faster and aid distance learning programs.
- It will increase my general information base and help me keep in touch with people and places.
- It will give us update of happenings around the world and would also reduce time spent at work in sourcing for data or information.

Communication was the third major theme in the expectations of respondents of the new technologies. Access to communication may be taken for granted in countries such as Bermuda with 89.52 main (fixed) telephone lines per 100 inhabitants (the highest teledensity in the world in 2007, according to ITU, July 2008). But in Nigeria, even with NCC's combined (main lines and mobile phones) teledensity of 38.09 per 100 inhabitants, ability to communicate is very important. As has been seen in the usage of specific ICTs, respondents use the telephone (mobile and fixed) mostly for (personal) communication within the country and e-mailing for international communication, though the rate of in-country e-mailing is rising. This was also reflected in the expectations of respondents about ICTs in the area of communication. There seemed to be a split between domestic and international communication with some respondents expressing their communication needs to be local as well as international. Many who believed that ICTs would facilitate communication for them expressed the need to stay in touch with friends and family living outside the country.

Emerging Issues: Awareness, Access, Affordability and Availability

Analysis of the responses in the open-ended section of the questionnaire as well as discussions and interviews with officials in both the public and private sector shows that four major factors would facilitate or hinder the rapid diffusion and usage of ICTs in Nigeria. These are awareness, access, affordability and availability, here referred to as "The 4 As of ICTs in Nigeria." The distinction between access and the last two "As" acknowledges the fact that these concepts are not interchangeable in a development country context. Indeed, this is why the notion of teledensity has been contested in its emphasis on numbers and failure to address internal inequalities. Also, in countries such as Nigeria, more than 65 percent of the population live in rural areas where the technologies are neither available nor affordable.

Awareness

Many respondents expressed concern about the lack of awareness of the "benefits of ICTs" especially among women, arguing that awareness (as well as the other As) will determine how rapidly ICT usage spreads in the country. They offered suggestions on ways of raising public awareness. Some samplers:

- (People should be made) aware of the fact that a lot of information can be passed and received both locally, internationally at minimal time.
- Introduction of computer studies in secondary and primary schools and teaching people at subsidized rate.
- Reduced price (cost), education through seminars such as "Educating people about how ICTs" will help people improve their business awareness, especially among students. If people are familiar with it in schools, they will use them in business.
- Public enlightenment on the advantages of ICTs— and what people stand to gain from the "venture." Motivation on the part of government, availability of resources.
- By making more people to be involved in the new ICTs in Nigeria because some literates are still not ICTs literate.
- ICTs can be rapidly promoted in Nigeria if the Primary Education Board can introduce it as a course at the primary, secondary and tertiary level. Also adult education program should be embarked upon to educate the masses on how to use the ICTs.
- Educate people on the use of the new ICTs. Make computer courses compulsory in schools. Provide more cyber cafés in towns and villages.
- Have schools where they (women and girls) can learn (ICT) use at a very cheap rate.
- Its practical use or application should be emphasized at all levels of educational sector in the country.
- There must be legislation to promote it, increase in government spending on science and technologies, training of staffs (government and private) and re-orientation of the populace on the usefulness and benefits of ICTs.

Access and Availability

It is one thing for Nigerians to be aware of the benefits of ICTs and another to have access to the technologies. Titi Omo-Ettu, a telecommunications engineer, defines access as the "distance a citizen must travel to be able to use the telephone and at a price he (sic) can afford" (Omo-Ettu, 2007). He differentiates access from teledensity, a measure he says does not work for developing countries because a high teledensity does not necessarily mean that majority of the people can afford a telephone. Stakeholders in Nigeria's ICT sector are convinced that universal access is attainable because of developments in the ICT sector in recent years, especially since 2001. Also, the ubiquity of the "umbrella people" and cyber cafés ensures that regular people, least those in the cities and towns, can easily meet basic their ICT needs.

Respondents suggested that government could expand access by subsidizing the cost of acquiring ICT training and the technologies themselves. They also said the government could get into the cyber café business and charge lower rates than are obtainable in commercial centers. Some of the interviewees said Internet Service Providers (ISPs) should charge lower rates so that cyber cafés can offer their services to the end users at more affordable rates. According to one major actor in the ICT sector:

> ISPs have to break even, I agree but it's not something that they will break even overnight by embarking on cutthroat prices. There should be no monopoly. We have to encourage as many ISPs as possible to go to each of the state capitals. Monopoly will eliminate competition and when there's no competition, it means there'll be a dictation to the people in terms of the price, the people are held to ransom by the solo ISP in the state concerned. There must be level playing field, competition, awareness, transparency and openness.

In the same vein, another major principal suggested that providers of the various ICT services in Nigeria should benefit from the huge population by reducing their costs to make their services more accessible. His argument was that population is a strength that the Nigerian ICT industry can exploit. For instance, if five million people have access to the telephone and can talk to another five million people, "it is a lot of empowerment. You begin to look at the traffic...if the investors provide these 10 million telephones, and we have to provide at affordable price to these people, then it is not going to take a long time for the returns to come" (Omo-Ettu, 2007). And if this happens, then the providers can offer their services at two Naira per minute because that is "what Nigerians can afford."

Affordability

The International Telecommunications Union (ITU) defines affordability as *"relative:* the cost of ... ownership should not exceed a certain percentage of family income" (ITU, 1998: 33). "Cost," "price" and "affordability" appeared constantly as a response to the research question: "In your opinion, what factors or conditions will promote the rapid diffusion and use of new ICTs in Nigeria?" While the cost of many ICTs fell drastically between 2001 and 2008, affordability remains a significant factor in the diffusion and use of ICTs in the country. For instance, as noted earlier, many of the respondents did their internet activity in cyber cafés. For many of them, each time they accessed the internet, they spent at least ₦100 for about an hour of access. Considering the fact that NYSC members earn less than ₦10,000 a month, with the lucky ones getting some accommodation and transportation allowances from their primary employers, there is an obvious economic constraint to the intensity and frequency of usage of many ICTs.[11] For instance, 86.2 percent of corps members who participated in the second wave of the survey reported that "cost" was an important factor that "would increase (their) ability to make calls more often on (their) GSM phone." Also, 36.2 percent of respondents said cost of access was an important factor that determined the frequency of their using the internet. In verbalizing their responses (and in the open-ended section of the questionnaire), respondents also reported that they would use other technologies more frequently if the costs of access were lower.

Conclusion

The responses of 408 questionnaire respondents are aimed at facilitating understanding of not only the practice of ICTs, but the issues that have emerged in the process of diffusion, usage and harnessing of these technologies as tools for socioeconomic development in the country. From participants' responses, one finds that factors such as awareness, affordability, availability and access are critical to the successful execution of the ICT4D project in Nigeria.

While the penetration level of many ICTs has increased in the country the diffusion is uneven and sparsely dispersed with higher levels of access and availability observed in major cities and towns. This distortion is not evident in the patterns of usage by the 408 respondents of the questionnaire because the gap between those who used or had access to the technologies and those who did not in the month prior to their participation in the research was not significant, relative to what was observed in the larger society, 60 percent of who live in homes without electricity. This limits the ability to generalize the ICT experiences of this group to the larger society, especially given the age, education level and the geographical location of the respondents and the size of the sample. Since they work and live in three major cities (that have infrastructures not available to most residents of other cities) research respondents are already positioned above the level of the "average Nigerian." Nevertheless, this group is a critical one for a number of reasons including some of those that make their usage pattern atypical of the general population: their level of education and general awareness. Therefore, how this group engages with these technologies, while not representative of 140 million Nigerians, does illuminate the general pattern of usage and diffusion in the country.

Notes

[1] The different branches of engineering, for example, were counted as one.

[2] The preponderance of female presence at various levels of the ICT sector in the country made it unnecessary to pursue a gender balance in the way I did in the first wave of the research.

[3] Many of these definitions were given in Chapter 1.

[4] While this extended the duration of completing the 16-page questionnaire, listening to the ways through which respondents rationalized their responses gave the researcher information that illuminated the issues much more than rigidly coded responses would have done. In the open-ended section of the questionnaire, the researcher probed for more complete responses and clarifications. However, there was no attempt to influence responses by offering suggestions or explanations.

[5] As the researcher read out the questions and recorded responses, each participant was given a copy of the questionnaire to read and follow during the administration progressed. This eliminated the problem of respondents not hearing or understanding the questions. In fact, many respondents would first listen to the researcher read the question, and then they would read it up for themselves. Others read along as the interviewer read each question.

[6] Many respondents would ask the researcher: "what does this one mean?" and the response usually was: "please answer according to *your* understanding."

[7] A follow-up question for those who said they had done an Internet search in the month prior to their participation in the research was: "What information were you searching for? (*Please give some of the keywords*)"

[8] Since the questionnaire was interviewer-administered, one was careful not to influence responses by interpreting or explaining the questionnaire once administration had begun. However, because of the process of verbalizations (referred to earlier), the interviewer was able to input the data as accurately as possible.

[9] This occurred in the few cases where attempts to explain the question to respondents were likely to result in an influenced response. In those cases, the interviewer recorded a question mark and moved on to the next question.

[10] Only very few of the respondents had direct internet access and therefore determining if an address was "outside the country" was problematic as most e-mail users in the country have Yahoo! and Hotmail accounts. Technically therefore their addresses are outside the country. However, the question was framed to emphasize the geographical location of the recipient or sender rather than the location of the internet hosts of their e-mail addresses.

[11] During the NYSC year, the federal government pays a monthly stipend to corps members. Primary employers in the places of primary assignment are only required to provide accommodation and transportation allowance. Many employers are unable to do this. Yet others offer supplementary stipends to attract corps members as they provide cheap (but knowledgeable) labor.

• CHAPTER SEVEN •

Potholes on the Information Superhighway...and Detours

Introduction

Poor state of roads is a common feature on the Nigerian landscape. This is especially so in the southern parts of the country where a combination of the heavy rainfall, awkward road networking, poor construction and maintenance, ill-maintained vehicles and reckless driving transforms the roads into death traps. Besides the appalling condition of the roads, regular checkpoints and blocks by all sorts of armed personnel (soldiers, police and thieves) also characterize the roads. These problems make some stretches of roads, especially the 1,000 odd kilometres between Lagos and Calabar (and through many eastern cities and towns) particularly dangerous. Those who travel these roads know they risk death by either road accidents, armed robbery or "accidental" or deliberate shooting by any of the groups of armed personnel at the numerous blocks along the road.

One of the expected benefits of information and communication technologies (ICTs), as articulated by many Nigerians is the potential for reduction in the need to travel. Telephones and the internet facilitate communication such that one does not have to travel physically to perform routine tasks. Besides voice communication, documents can also be easily transmitted (either through the facsimile machine or as e-mail attachments). Many Nigerians therefore expect that ICTs will build a virtual alternative to the roads in the country. Unlike the physical road, the virtual highway (or the information superhighway) is expected to lead directly to the global information society, and along the way, unlock access to immeasurable amount of data, information and knowledge.

Roadside shops, hawkers and service providers such as auto mechanics litter Nigerian roads such that people—especially those travelling over long distances—can shop for groceries and even home furniture on their journey from one point to the other. Similarly, on the virtual road, there are stores and shops where people can trade in all kinds of goods and services. The information superhighway is also a huge library, a university, a playground, a community hall and other imagined places. Nigerians just need to acquire the technologies as quickly and as much as possible in order to be part of the global information (or network) society. These are the contents of the narratives about ICTs and their capacities to lift Nigeria from socioeconomic obscurity to prominence—to make the country "globally competitive," according to the policy on ICTs. The reality however is that just as the Nigerian road is filled with potholes and checkpoints, so is the virtual highway. These hurdles present certain challenges to Nigeria's ability to navigate the information superhighway smoothly and rapidly (thus enabling leapfrogging from the country's peripheral capitalist economy to post-industrialization).

Given these expectations about the possibilities of ICTs for socioeconomic development in Nigeria, it becomes imperative that one examines the factors that are likely to mediate the successful harnessing of ICTs for socioeconomic development in Nigeria. These factors are here referred to as "potholes" while "detours" describe measures and coping mechanisms that policymakers and other stakeholders in the ICT industry have devised to circumvent the obstacles. The relevant potholes in the present context are the institutional framework, state of the infrastructure, poverty and illiteracy, cultural framework and ethnicity.

Institutional Framework

Van de Ven has argued that "innovations not only adapt to existing organizational and industrial arrangements, but they also transform the structure and practice of these environments" (cited in Montealegre, 1999: 199). For this transformation to occur there has to be an appropriate framework on which the innovations must build. Until 2000, when the National Telecommunications Policy (NTP) was released, there was no policy—or "appropriate framework"—on ICTs in Nigeria. And a year later a specific policy on ICTs—the National Policy on Information Technology (NPIT)—was released to address some of the shortcomings of the NTP in offering a comprehensive coverage to the sector.[1] Together, these two policies provided the starting point for the development of ICTs in Nigeria. They also signalled to Nigerians and the world that the government accorded adequate priority to ICTs and believed in their potentials to stimulate the country's economy. The policies created the needed institutional and legal framework for innovation in ICTs. Despite these efforts, the nature of the Nigerian society raises important issues regarding the prospects for successful accomplishment of ICT4D goals in the country. Stakeholders express concerns over the process of implementing the policies and the adequacy of the legal framework to circumvent the factors that hindered achievement of goals of previous national development plans. (See Chapter 3 for an overview of the evolution of development plans in Nigeria.)

While many in the IT industry enthuse about the wonders of e-commerce and how this form of business will literally advance a world of economic opportunities for Nigerian businesses and experts. As one interviewee noted, in the digital revolution, the commodities of value are no longer material but information and knowledge and these can be traded regardless of geographical boundaries. And as many in the Nigerian ICT industry argue, even physical goods can also be sold to consumers in other countries without human physical movement. These transactions will be enabled through electronic commerce (e-commerce) and mobile commerce (m-commerce, which refers to business transactions done over the cellular phone). And thus, as stated in a previous chapter, Nigerian newspaper headlines are filled with "e-" to prefix the promise of the digital revolution in the Nigerian landscape. But a major obstacle to the achievement of these e-goals is the absence of an appropriate legal framework that clearly delineates the rules of business and at the same time protects merchants, customers, and investors. Many foreign businesses often complain that the cost of doing business in Nigeria is higher than in other countries with the same level of development, with most of that cost being "hidden," a code for bribes and other unofficial expenses that determine the success of many ventures in the country. Beyond these, there is the high incidence of fraudulent activities associated with doing business in Nigeria.

In the 1990s, the Nigerian government awakened to a scam (advance fee fraud) that some Nigerians were unleashing on arguably greedy and gullible foreigners. In one scenario, someone in Nigeria would send a letter or fax to another person in a foreign country informing the addressee of some "forgotten" or "unclaimed" money in the Central Bank of Nigeria. The details varied but it was usually about how a company that had gone out of business or an individual that had long been deceased was being owed large sums of money by the Nigerian government for work the company or individual had done for a previous administration. The money was now ready to be paid but the company or person(s) no longer existed. However, the addressee could pretend to be the deceased or no-longer-existing entity and claim the money with the assistance of the addresser who would handle the red tape for a share of the loot, usually an amount as "little" as 10%. As a first step in the process, the foreign recipient of the scam letter would be asked to remit some money as transaction fees[2] as well as give particulars of the bank account in his or her country to which the money in Nigeria would be transferred. When the foreign recipient paid out the first instalment, there were usually more requests for additional payments because of some obstacles in processing the transfer. In the end, the recipient would have exhausted his or her life savings and in many cases—as a couple in Vancouver, Canada did—take out bank loans to send to Nigeria in anticipation of millions of dollars (of unearned money) that would be transferred into their accounts. The situation became so ugly and earned Nigeria such a poor international image that the Babangida Administration inserted into the Nigerian Penal Code a law that became known as 419 (Article 4, subsection 19 of the penal code), which also assumed the name for the scam itself (Four-One-Nine).[3]

The scam has been widely reported in the foreign media and many foreign affairs departments have issued "travel alerts" warning their citizens against doing business with Nigerians unless their credentials were authenticated by foreign embassies and high commissions in Nigeria. These publicity and alerts have not deterred those Nigerians engaged in 419 since there are always people eager to reap where they did not sow, but they have made many foreigners wary of doing business with Nigerians. This is particularly so now that 419 has gone online. Many "419 groups" (also known as "Yahoo boys") in Nigeria now daily send e-mail to people overseas peddling the same story that they did in the days before they had access to the internet. In a digital world, access to names and particulars of potential 419 targets is even easier than in the past.

While waiting to access the internet at a cyber café in Ikeja, Lagos, I observed an e-mail being composed by a young man claiming to be a lawyer whose firm had been awarded a contract to negotiate for the payment of a huge amount of money to a foreign company that had done business with the Sani Abacha government (1993-1998). He was looking for a foreign legal firm to partner with his own firm to execute the project. It did not occur to the person sending out the e-mail that the recipient (presumably a lawyer) would realize that any foreign firm that was endowed enough to execute a contract of that magnitude would have already had its own legal team to negotiate on its behalf. But it did not matter if this particular e-mail recipient responded or trashed the e-mail. The "Nigerian lawyer" knew that out of probably 50 such e-mail messages sent out, one person would be greedy enough to swallow the bait. And so "419" thrives, causing enormous problems for those Nigerians engaged in legitimate businesses and looking for foreign markets and investors

This presents a challenge in the way of Nigeria's achievement of global competitiveness through ICTs. As Oyawoye, special adviser on IT to former President Olusegun Obasanjo, points out: "There is a lot of money outside looking for (underserved) places (such as Nigeria) to go. But...because of what (foreign investors) have heard about how difficult it is to do business in Nigeria and how much government is involved in everything," Nigeria is one of the countries with the least foreign direct investment (FDI).[4] In 2000, the country had a foreign direct investment (net inflows) of US$1.1 billion though compared with the Sub-Saharan Africa total for the period (US$7.3 billion), the country did not fare too badly.[5] It did even better as the most recent estimates available for 2006 shows: a total FDI of US$31.66 billion (World Factbook, 2008). This represents a net inflow of 4.7% of the GDP for 2006, according to the World Bank (2008).

Another concern about business transactions, especially online, with and by Nigerians is that Nigeria is essentially a cash society. Everyone pays cash for everything—from the purchase of peanuts to a brand new car. Use of credit and debit cards has only slowly begun but too few people may be eligible for "plastics" to make that medium of business transaction routine in the country. Also, the credit cards are, in many cases, debit cards because customers are issued cards that are directly linked to a checking or savings account of the card-issuing bank, and transactions are instantly debited to their bank accounts.

Cheque transactions are also limited since there are no guarantees of sufficient funds in the bank account of the issuer. Without an officially sanctioned penalty for "bounced" (non-sufficient-funds) cheques, these financial instruments are considered highly suspect. Also, there is no framework for debt collection in Nigeria, as obtained in the developed countries, which would ensure that people are forced to honor their financial commitments and automatically penalized for not doing so. In the past, cheques were honored with certain restrictions. For instance, a "crossed cheque" which could only be paid into a bank account (and not exchanged for cash) and certified cheques had value that have since deteriorated, following the crisis that beset the banking industry in the mid-1990s. In a few cases where cheques are accepted, parties to the transaction meet at the bank and transfer the money to the recipient's bank account before the deal is concluded. This ensures that the goods and services are actually paid for. It also eliminates the need to carry huge amounts of cash and thus expose oneself to armed robbery attacks especially in places such as Lagos and Port Harcourt where the incidence is very high. But it is a time-consuming and cumbersome process.

State of the Infrastructure

Telecommunication infrastructure: In a 2002 United Nations Economic Commission for Africa (UNECA) report of a survey on the competitiveness of the business environment of 24 African countries, Nigeria ranked constantly at the bottom or near the bottom in all categories. The country performed particularly poorly on roads, ports, railways and airports. It also ranked 20th on the quality of telecommunications (UNECA, 2002). While there have not been noticeable changes in the general infrastructure—and it can be argued that they have gotten worse—the telecommunications landscape has changed significantly as a consequence of the entry of digital mobile phones in the country. In 2007, the government regulator, the NCC improved on its performance in this area by approving the unified licenses to existing and new providers. By the end of that year, there were 12 providers of mobile, fixed and fixed wireless phone services in the country. There has also been similar growth in other sub-sectors such as the diffusion of broad band internet access, VSATs, a local internet exchange protocol (that enables local routing of internet traffic) and increasing manufactures in computer hardware and peripherals. However, as innovative as ICTs are, they are not independent of the basic infrastructure such as electricity, the state of which continues to be worrisome even to ICT enthusiasts in the country.

Furthermore, the rapid growth in the diffusion of the relatively affordable GSM services has resulted in congested networks such that subscribers to one network have difficulties connecting to subscribers in other networks creating multiple subscriptions to different networks. This is particularly challenging in the rural areas where everyone is familiar with the word "network" which has become shorthand for everything that hinders a user from making a phone call. Boye Olusanya, formerly head of the customer services division of Econet (Celtel, and now Zain)—one of the digital mobile phone providers—argued that the problem would be eliminated when more people subscribed to the GSM networks and created more mobile-to-mobile calls thereby reducing the dependency on Nitel and the necessity for GSM calls routed via the public switched network. By 2008, there were obviously more mobile-to-mobile calls, fixed wireless connections and a second national carrier (besides Nitel). Yet, the problem of congested networks, especially across networks, remained. Interconnectivity has therefore remained a major problem in the expanding mobile telephony industry in Nigeria.

Power infrastructure: Unreliable electricity supply has been the bane of businesses and Nigerians in general due to both low production and ineffective distribution. In 2005, 23,599 million kilo watts of electricity was produced in the country, far below the production level for Sub-Saharan Africa, 349,000 million, and 242,924 million for South Africa (World Bank, 2008). Fewer than 10% of the 51% population who live in the rural areas have access to electricity. In the towns and cities where there is electricity, its presence is felt more in its absence leading to the nicknaming of the public utility company, National Electric Power Authority (NEPA), as Never Expect Power Always. The Obasanjo Administration promised a legacy of uninterrupted power supply. Instead, he gave Nigerians a new name for the same problem. Following reforms in 2005, the agency was renamed Power Holding Company of Nigeria (PHCN). The new name has not solved the problem of unreliable electricity and Nigerians have therefore nicknamed the new company as Problem Has Changed Name (PHCN).

Besides the low production of electricity, another major obstacle that besets power supply in the country is the incidence of theft and vandalism of power infrastructure. Often, power lines and meter boxes are stolen and resold by people colluding with PHCN staff. It is common to see meter boxes mounted outside homes all fenced up to protect from theft. Other times, PHCN properties are vandalized in places where indigenes hope to pressure the government to accede to their demands. In most cases, government continues to be indifferent to their needs and the people have only succeeded in worsening the already poor state of electricity in the country. As one interviewer put it:

> You put fibre wire there and someone goes and steals it. This wire is almost useless unless you know what you want to do with it. Just because the thing is passing through their land, they think they can use that as leverage so that government can come and put a school in their village.

Electricity generation and distribution undoubtedly affect the diffusion levels of ICTs in Nigeria. While mobile phone services are now affordable they are still very expensive relative to the income of the average wage earner. The gross national income per capita in 2006 was US$620 (World Bank, 2008), less than two dollars a day. This greatly impacts the ability of many Nigerians to pay for the relatively high cost of ICT services and products. Providers of these services and products attribute the high costs of their services to the extremely high overheads of business in the country especially given the infamously epileptic nature of power supply in the country. According to Olusanya, a mobile phone service provider needs at least two generating sets at each base station. One generator is a back up for the public power supply and the second generator acts as a back up to the first set—all to ensure uninterrupted service.

> Infrastructure in the country has been a big (problem). In North America, when you talk of cell sites and things like that, you don't worry about generator. Here, you need a generator and a back-up generator and if you are not careful, you need a back up to back-up generator.

Emmanuel Ekuwem, the national president of the Association of Telecom Companies of Nigeria (ATCON), expressed similar concern: "Energy is an important factor in this revolution. If not probably handled, the (information and communication) revolution will fail. It's energy that drives the machines." Ernest Ndukwe, the chief executive of NCC, echoed Ekuwem's sentiments, adding that electricity plays a critical role because "without it, it's very tough. I think this is probably the biggest threat to the expansion of ICT use in the country."

Import dependency: Another general infrastructural obstacle lies in the fact that until very recently, Nigeria did not produce any ICTs. While some of these technologies are produced or improvised locally, the bulk of ICTs are imported as finished products. The consequence is the overwhelming dependence of the ICT industry has on importation of products and services. Ekuwem agreed that "undue reliance on foreign companies" will severely "upset the successful implementation of the revolution." He and other IT practitioners in the country are therefore insisting that the government, while acknowledging the globalized nature of current national economies, should not abandon the development of ICTs in the country to foreign companies.

> Some of the jobs should be done by Nigerian companies. After all, we are not making routers and switches and Ethernet cards and microprocessor chips and motherboards. Those ones can be imported, but when it comes to assembly, systems integration, turn-key projects implementation, we have sufficient local manpower to handle such projects. Of course when we allow local companies to handle such projects, you are also imparting skills on the job, or auto imparting of skills via hands on…So if they decide to give all the contracts to foreign companies to execute, local companies will feel disgruntled and discouraged and sidelined. This will have a negative social impact on the whole revolution.[6]

According to an official in the Federal Ministry of Information and Communications, the lack of basic infrastructure as well as the import dependency will stymie the country's ability to achieve its ICT-related goals. But this does not have to be so, she argued, because Nigeria has the capacity to locally produce ICTs.

> If you go anywhere in the world where (ICTs are produced), Nigerians are there driving these things…in the factories, development team and research. It's possible for them to come back to Nigeria and do the same things here. But they can only come when they know that they can go to the bank and get their money in minutes, and not when they have to take a bed to the bank before they get their money. And not when they are being harassed by armed robbers everywhere. And not when they get to government and will have to pay half their contract price to people here and there be-

fore they get the contract...If corruption is taken away...many more people will get involved. In summary, we have the capability.

She thus makes a connection between the need to be self-reliant in the ICT sector with other problems of development in the country such as insecurity of life and property (crime), poverty and corruption, as identified by Yesufu, 1996.[7] Interestingly, the level of poverty in the country has also been blamed for the high incidence of armed robbery because, as one interviewee argued, Nigerian youths are desperate as a result of lack of employment, the rate of inflation, and living side by side with rank opulence. But while ICTs are expected to create employment, wealth and alleviate poverty, these very conditions may hinder the development of the ICT industry. As official from the Federal Ministry of Information and Communications pointed out, if ICT-skilled Nigerians in the Diaspora are not comfortable enough to return to the country, the development of ICTs will continue to depend on external conditions, and this dependency is not sustainable. Corruption is also another problem that deters many Nigerian ICT experts abroad from returning to work in Nigeria. Again, ICTs are expected to solve this problem by making transactions transparent and information so available that one would not have to bribe anyone to access government services. It may be a long time yet before Nigeria reaches this level of information transparency and availability. Currently key government functionaries and private stakeholders working in the ICT sector are unresponsive to requests for information. Many do not respond to e-mail messages. In this respect, the expressed expectations about the capacities of ICTs are yet to connect with the colonial work habits of Nigerian bureaucrats where government information is perpetually "secret" and "classified." I return to this point later in the chapter.

Infrastructure Detours

Many Nigerians are nonetheless optimistic that the country has the capacity to scale the hurdles and potholes presented by the current levels of ICT infrastructure in the country. In many cases, they have found and improvised detours and short cuts in the journey toward the global information society. Three of these detours are discussed in the following paragraphs. As ICTs are integrated technologies and often depend on other technologies, these detours are not isolated developments but are mutually dependent. There are therefore inevitable overlaps in the different areas. The particular strategies of focus are: mobile telephony as detour for inadequate traditional telecommunication infrastructure, alternative sources of power to circumvent unreliable public power supply, and local manufacturing and improvisation of ICTs to reduce import dependency.

Mobile telephony: The commencement in August 2001 of cellular phone services radically changed the telephony landscape in the country. As at June 2008, there were 51 million active cellular and 1.6 million fixed/fixed wireless phone lines in the country, raising the country's density to 38.09, a growth that has been described as phenomenal (NCC, 2008). The rise in the number of cellular phone subscription between 2001 and 2008 resulted from several factors. First, on the surface, the technology does not depend on Nigeria's preexisting telecommunications system. This means that Nigerians without core infrastructure such as telephone main lines or electricity could still access a GSM-enabled cellular phone service. The only conditions for a phone line are the finance and residence in an area of coverage. Second, and related to the first, the pre-paid packages offered by the service providers allowed even for the homeless to afford a telephone—not in terms of the price, but because one did not need to connect wires, show evidence of regular income or make contractual commitments. A reference by a stakeholder to beggars (the North American equivalent of homeless street panhandlers) owning cell phones was an exaggeration of the reality just three months after the digital mobile service providers rolled out their services. However, by 2008, it was common to see beggars on the streets of Lagos with cell phones. Also like many countries in Europe (and unlike North America), Nigerian cell phone users pay only for the calls they initiate. In-coming calls are at the originator's cost. That way, users can manage their credits and tailor usage to meet their ability to buy the "re-charge" (pre-paid phone) cards. Though only less than 30% of Nigeria's population is part of the "cell phone world," the diffusion of cell phone usage has been relatively broad-based, cutting across gender, age, education and socioeconomic levels. There are expectations that ultimately, cellular phone service would also expand much further into the rural areas particularly with the increase in the number of providers and diversity of the technologies.

Alternative sources of power: Manufacturing and sale of generators and their parts and repairs, power back ups and uninterrupted power supply (UPS) systems are among the fastest growing businesses in Lagos and some other big cities in the country. In the ICT industry, when someone says he (rarely she) wants to buy a computer system, what he means is a computer, printer and UPS. In all the places visited during the research, all the "systems" I saw included a UPS on the floor under the computer stand. The average would be two and in some places—depending on the system's function—there might be three UPS units, with each providing back up for the others. At cyber cafés, all the computer terminals had a minimum of one UPS that might provide up to 20 minutes of power when the public power supply was interrupted. In every office (besides government offices), there was at least one high-capacity generator with automatic power switches often with few seconds of lag time after power interruption. In many of the cyber cafés and business centres, customers were charged higher prices for services if a generator was providing the power. This was very common in cities that had few service providers, therefore leaving the customer with fewer options.

Poverty and Illiteracy

About 66% of the Nigerian population live in poverty. The adult literacy rate in 2003 was 69.1% (World Bank, 2008). Poverty and illiteracy are therefore two major socioeconomic problems in Nigeria. The issues are related because as studies have shown, access to education empowers the individual by giving him or her necessary tools to have healthy, productive and dignified life. In Nigeria education was always a fast track to wealth and privilege—at least before the high graduate unemployment rate seemed to indicate that formal education was a waste of productive years (years that could have been better spent on learning a trade or apprenticing to become a trader). The obstacles posed by poverty and illiteracy to the goals of harnessing ICTs for socio-economic development cannot be overstated.

Ideological and Cultural Framework[8]

It is widely argued that technology is not just the equipment, but comes bundled with certain sets of ideas. For instance, in feminist analysis of women and technology, the point is frequently made about the ideological context at the point of the invention of technology. It has been argued that the computer was designed with a certain male notion of rationality and efficiency, and the internet is specifically a creation of an institution of male supremacy (the United States military) and therefore comes appropriately bundled with male ideologies that particularly exclude women. As Light (1995: 134) notes, the "histories of computer networking document the evolution of electronic communication in a male-dominated environment" with the first computer network, ARPANET, evolving in the 1960s out of the US military. According to Spender (1997), given the origins of computer technology, men got in first, shut the door behind them and re-arranged it (the technology) to suit their needs. The consequence is that in many cases where women have equal access to the technologies, they are restrained by the fear that they are stepping into male territory. Agreeing that "technology does not develop independently, but is part of a particular social-economic and cultural set up which circumscribes relevance and meaning," Zoonen (1992: 14), however presents a contrary and constructivist analysis to explain technology as a discursive practice. She argues that the "meaning and social significance of technology is not pre-given but established in ongoing historically and culturally specific discursive practices." (p.12). Bush (1983: 165, cited in Zoonen, 1992) also argues that the meaning and relevance of technology for everyday life are constructed in the context of production, usage, environment and culture.

The research revealed that many Nigerian users of ICTs are not conscious of any foreign ideological contents in the technologies. Those who make reference to the overwhelming importation of ICTs do so only within the context of the impact of dependency on foreign markets on the economy and the sustainability of the ICT project. Usage of ICTs in Nigeria constitutes a set of discursive practices and their meanings and social significance are not pre-given. To further strip ICTs of their "foreign ideologies," it was found during the research that, ICTs—especially the internet—are mostly used within the country to solve locally generated problems. One notes in passing that this seemingly contradicts official conceptions of ICTs as tools that will help the country to be globally competitive—unless one argues that in a globalized era the local is global just as the global has become local.

The debates on the ideological nature of technologies are relevant in the Nigerian context only as they highlight the conflict between certain notions of the information society and prevailing practices in the Nigerian society. Examples of this conflict include the lack of openness and the mystification of knowledge that characterize the Nigerian ICT landscape. The civil service (and other areas of society) has a certain culture of secrecy and suppression of information. Years of military rule following shortly on the heels of colonialism have created a people distrustful of each other and withholding information in apprehension of reprisals. Also, the ethnically and religiously divisive nature of the society reinforces an ideology that contradicts the freedom of expression that is entrenched in the Nigerian 1999 constitution. Furthermore, given the low literacy levels and the disparity between those who have access to education and those who do not, knowledge has often been mystified at many levels. For instance, for a woman to trade in the market, she not only goes through the official process of acquiring a stall (or space) but also must be "initiated" into the association (sometimes operating as a cult or pressure group) of those already selling the same products in the market. Initiation into the group guarantees a peaceful business environment for the newcomer as well as access to the secrets of the trade—such as where to acquire inexpensive products to maximize profits.

'Official Secrets' Versus Open Information Society

Old bureaucratic and colonial ideas about "Official Secrets" still shape thinking about ICTs in Nigeria. Access to policymakers and documents in the ICT sector remains restricted and while many public agencies and departments have expended millions of Naira commissioning and launching websites, useful information is scarce and often contradictory. Responses to electronic queries are at best minimal.[9] The absence of information from official sources contradicts the idea of the information society with its assumptions about openness and transparency—or something Friedman (2002) might call the democratization of information. In this environment, the colonial and military ideologies of secrecy and disconnection from the people that have beleaguered the Nigerian government for years constrain the ways in which Nigeria can compete globally using information technologies.

Information has always been important to governments in Nigeria. The Ministry of Information and Communications at the federal, state and local government levels remains an essential government institution even in lean times when some ministries are merged or abolished. But institutional information in Nigeria has also been unidirectional with little or no forum for feedback. In the country, information flows *from* the government *to* the people through several media channels. This unidirectional flow was especially reinforced because until the 1990s, all the electronic media were fully owned and controlled by the state. The federal government owned and controlled two national newspapers and every state government had its own newspaper. During the Babanginda Administration (1985-1993), a program was established with the task of mobilizing the people. The concept of mobilization implies a two-way communication process. But MAMSER operated within the military framework of the day: information flowed in one direction. In the current context of "information society," information flow must necessarily be a two-way process. But so far, while Nigeria exalts the importance of information for development, the understanding of information flow is still embedded in the unidirectional orientation of colonial and military-era logic.

Monopoly on Knowledge

There is a certain mystification of knowledge about ICTs that clashes with policy goals of general IT literacy and universal access to achieve overall socioeconomic goals. Access to knowledge of ICTs in Nigeria has raised new sources of power and class of people who consider their knowledge of and access to ICTs as a status symbol. This creates a monopoly for those who parade as "IT persons" and withhold their knowledge and information in trepidation of losing the new source of economic and social power. Oyawoye finds this "very silly" because with access to ICTs, anybody can easily acquire information about anything on the internet.

> It's all part of this misguided elitist thing. People think that acquiring computer skills is something that only a few people should have access to, and so they set up very elaborate training institutions and carry huge overheads and therefore charge very high fees. Some actually set up elaborate executive schools for top executives and children of very rich people...

The mystification about ICTs is further exacerbated by the cost of training and acquisition of the technologies. As Oyawoye said, it is very easy to learn how to use a computer when one has access to it and thus bypass the need to pay exorbitant fees to "computer schools" which barely offers adequate hands-on training because of the high ratio of students to the technology. It would be easier and cheaper for people to learn how to use the computers if they had the systems at home, said Oyawoye. He criticized the media for not helping to educate people about ICTs because journalists are preoccupied with "putting their opinions or trying to make and break politicians." This leaves the field open for "anyone who has some knowledge about anything to put a premium on that knowledge and mystify it...There is a lot of ignorance passed on by journalists who don't even know what they are writing about."[10]

The realization that there is wealth in a knowledge-based economy is interpreted by many IT-aware Nigerians as a monopoly on and commodification of knowledge. The Nigerian environment makes this especially attractive because of the phenomenon of contracting—or in the language of ICT age, consultancy or outsourcing. People hoard what they know because they believe that if their knowledge is dispensed as public service, everybody would become an expert thus cheapening their services or rendering them irrelevant. It is a common saying in Nigeria that if everybody knew how to repair cars, then auto mechanics (or service technicians) would be out of business. In Nigeria, this assumes an added meaning because auto mechanics particularly prevent their customers from trying to understand the problems with the cars to avoid a do-it-yourself the next time the car has similar problems. Also, it is common for commercial drivers (bus and taxi drivers) to deride women or elderly drivers (or anyone who looks like an "Oga") with comments such as "I go drive myself" or "You fit buy car but you no fit pay driver."[11]

It may not be obvious to modernization theorists—and other Weberians who argue that a major indicator of modernity is specialization of functions and roles—but Nigeria is a society driven by a certain notion of specialization and clearly defined occupational boundary lines. This contrasts with an information-society ideology of multi-tasking (or jack-of-many-trades-and-master-of-all) in which anyone with training in any background can do anything because ICTs facilitate the process. Indeed in a society such as Nigeria, where ICT usage is not yet pervasive, many people currently in the field had no specific ICT training. But they have come into the sector and created associations aimed at excluding others through rigid eligibility rules, as well as certification and licensing procedures that ultimately restrict entry into the field (though the associations frame these rules in the language of standards and excellence).

Ethnicity as Pothole

One of the features of the information society (or capacities of ICTs) is the compression of time and space. Many Nigerians speak longingly about the ways in which ICTs would reduce their need to travel or physically conduct businesses. And given the poor condition of roads in Nigeria, the spatio-temporal compression enabled by ICTs would benefit many. But when the argument about the collapse of time and space is extended, it appears that the effects in Nigeria will be contradictory. In the country, unity has historically been fragile and political actions are carefully managed (or manipulated in some cases) to avoid any hint of ethnic, regional or religious biases. In Nigeria, space is literally a contested terrain. There are constant ethnic and religious conflicts rooted in claims to space (or land). One comes from a specific, timeless and static "place of origin" regardless of place of birth or residence. A child born in Lagos to parents who were born in Owerri will never be a Lagosian (belonging to the Yoruba ethnic group) even if he or she spends the rest of his or her life in Lagos and has never lived in Owerri. This individual will always be identified with Owerri (forever designated as Igbo) and will engage with the political and socioeconomic life of Lagos as an Igbo, and therefore a "non-indigene."

Employment, university admissions, political appointments and other types of political patronage are based on "state of origin." Thus the current federal government in Nigeria has 36 members in the federal cabinet—each representing the 36 states of the federation—to reflect ethnic balance. Overarching these issues is the Federal Character principle, which essentially refers to a deliberate attempt to reflect the diversity of the country in political appointments and employment in federal government departments. The principle was adopted as a post-war national policy by the government of General Yakubu Gowon in 1973 to assuage feelings of marginalization in a country of many previously independent and distinct nations. While this principle has not reduced the spate of ethnic and religious crises in the country, its existence ostensibly mitigates many conflicts by reducing the perceptions of marginalization. This basic need to forge unity in a country where the different ethnic groups would rather split and assert political autonomy seems diametrically opposed to some fundamental features of the information society such as integration and elimination of spatial boundaries and hierarchies.

Some have argued that the diffusion of ICTs and the shift from a resource-based economy to a digital economy will positively rearrange the country's geopolitical configurations. According to Nosa Omoigui:

> Given our history of ethnic fissures and mistrust, the Information Age would have major geopolitical implications on Nigeria. I predict that in the 21st century, intellectual capital, not mineral resources like oil, would be the most strategic and valuable global resource. The impact of this phenomenon on Nigeria cannot be overestimated. It suggests that as time goes on, a country's geographic placement and mineral resource endowments would take on less and less significance. In contrast, the skill-level and educational competence of its workforce would become the primary determinants of economic prominence.[12]

The country is endowed with mineral resources, the main one being oil. The federal government controls all natural resources, and manages the allocation of revenue derived from them. Over the years, this has resulted in accusations of marginalization especially by the oil producing states of the southern Niger Delta region whose people have often expressed the sentiment that their resources are being exploited for the development of the North—a non-oil producing region. The frustrations of the people of the Niger Delta derive not only from the fact that they have no control over their natural resources, but also that they are suffering from the unmitigated effects of oil exploration activities on their land and waters. These activities have over the years rendered much of this area useless for other forms of economic activity. The geopolitics of oil in Nigeria, the subject of much scholarly research and writing, peaked during the Abacha years with the state execution of the famous Ogoni Nine—environmental activists from Ogoni, an oil producing area in Rivers State. The killing of the Ogoni Nine (which included international author and playwright, Ken Saro-Wiwa) resulted in the expulsion of Nigeria from the Commonwealth and other sanctions by countries such as Canada and the United States against Nigeria.

Ethnic Detours

When the Obasanjo Administration assumed power in 1999, two measures were introduced to address the sense of marginalization expressed by people in oil-producing areas. The Niger Delta Development Commission (NNDC) was a program aimed at dealing "urgently and fundamentally with the developmental needs of the Niger Delta and bring sustainable prosperity and peace to the area...and sustainable development" to the region.[13] Also, the federal revenue allocation formula was rearranged so that three percent derivative revenues went to oil producing states—in addition to their regular revenue. However, as in many of the administration's projects, these initiatives failed to achieve their objectives—even though for about two years, oil-producing states were receiving 13 percent more than other states from the federal purse. It is not apparent that the NNDC has achieved its goal of facilitating the development of the oil-producing states, or that the additional resources met the needs of the majority of the people in the region. Also, in 2002, the derivative revenue was suspended following a Supreme Court ruling on a different but related case.

Information technology enthusiasts such as Omoigui believe that the politics of ethnicity and resources will become less important in a digital economy. In the new economy, the commodity of value will no longer be geographically based but virtual and "brain power" will signal the difference between those who succeed and those who do not. Similarly, access to the seaport—as in many southern states—will not be significant in determining the economic prowess of each of the regions in the country. Lagos has always had a developmental advantage because of the presence of Nigeria's two major and busiest seaports (Tin Can and Apapa) on its coastal line. But in the new economy, it would no longer matter, the argument goes, and therefore ethnicity, often constructed around place of origin, will not hinder the application of ICTs toward developmental objectives.

> Proximity to the Atlantic—access to seaports—would no longer be an assurance of commercial power. The new seaport of the 21st century would be the handheld device, the personal computer, the television. Arguably, the primary export would be brainpower and locally authored software services hosted on web-sites. The ports from which these products would get dispatched would be Internet access nodes. Fiber-optic lines and the airwaves would complement the oceans as the primary conduits of international trade.[14]

Indeed, a deputy director and head of the technical services division in the Ministry of Communication, believes that ICTs—at least their utility as communication tools—can build understanding and unity in the country in ways that conform with the notions of the information society as a space in which everyone has access to the same information and knowledge.

> It's (all about) communication and if you are able to communicate effectively with each other, it will help you know what is happening. A lot of our problems come from not knowing and a lot of suspicion we have of one another. ... For instance, I spoke with someone from Bauchi who comes from a rural community (that is) trying to have a telecentre. One of the (motivations for setting up the centre) was that in the last (ethnic) riot in Lagos (between some Hausa and Yoruba groups), it was difficult for them to really know what was happening. The (people of the Bauchi rural community were) just depending on rumours and you cannot depend on rumours to really know if your people are really being killed. So if there's free information flow, it will defuse some of these tensions that we have and we'll really know what is happening.

It is still too early to assess the possibility that building an information society in Nigeria will defuse any tension in the country. Indeed, there is concern that the opportunities for communication and dissemination of information may just be another tool in the hands of those who exploit religion and ethnicity to achieve personal and political goals in the country.[15]

Conclusion

Generally, many Nigerians do acknowledge that there are many potholes such as the state of the infrastructure on the country's access route to the information superhighway. There is, however, overwhelming optimism among those interviewed that the detours and other improvisations will lead to a successful completion of the journey to the global network society. Ekuwem suggested that in the near future all the local government areas in the country would have access to the internet because of the number of applications for licenses to provide telecommunications services pending at the NCC. "And the NCC is giving a lot of priority to those wanting to go to the rural areas... The moment you put a BBB or VSAT node in Abak (a local government area), with VOIP (Voice Over Internet Protocol), a gateway interface with the public telephone network of Nitel or a PTO..."[16] the technologies will multiply with usage and spin off other applications and economic activities. Ndukwe of the NCC went as far as predicting that "a year from today one will not even recognize the Nigerian telecommunications landscape."[17] He pointed to emerging developments in the sector such as the entrepreneurial spirit of Nigerians who are jumping in following the lure of the enormous earning capacity of those already in the sector.

These optimistic remarks are undoubtedly well founded. At the policy level, it would appear that the appropriate institutional framework has been developed through the formulation of policies and establishment of implementing agencies, and the proposed bill to set out legal guiding principles for the sector. But interviewees' optimism probably greatly underestimated the depth of the potholes. As Titi Omo-Ettu declared: "E-business thrives on law; e-commerce on trust." It will take more than official policies and time to establish trust between Nigerians and between them and foreigners wishing to do legitimate online transactions in the country. For starters, it will be extremely difficult for Nigerians to "buy and sell on the internet" without the availability of trusted instruments for non-cash transactions that have built-in mechanisms to protect parties to the transaction from fraudulent practices.

In another area, the various measures to ease the public power supply problem have generally worked for those who can afford power generating sets and UPS. But these private solutions are obviously very temporary and geographically limiting. The use of UPS and generators presupposes an occasional presence of public power supply. Even then, the usage of generators is not sustainable for business purposes because of the exorbitant cost of fuelling and maintaining them. This is why those who utilize them for private purposes switch on their generating sets only at nights or on special occasions. In the rural areas where public power supply is non-existent, those who can afford generators use them only at night or during important events such as weddings and funerals. Thus, the success of the diffusion of ICTs and their usage as tools for socioeconomic development cannot be achieved through private provision of electricity. Until the general power supply is stabilized, as well as expanded to the rural areas, electricity is likely to remain a key hindrance in the development and diffusion of ICTs in Nigeria.

In conclusion, one argues that while great strides have been made particularly in the area of mobile telephony and local improvisations in the years since the policy on telecommunication was released, the journey ahead is as difficult and the terrain as treacherous as what occurs on physical roads in Nigeria. Often, at the onset of the rainy season, public works departments fill over the potholes with gravels. At a more extensive level, government contracts are awarded for road repairs and constructions only for the contractors to compound the problems by littering the roads with mounds of laterite, gravels, sand and construction equipment which hold up just long enough for the contractors to collect their fees. Further work is abandoned and the roads are left in a worse state than before. Also, on many roads in southern Nigeria, it is common to find young jobless men digging up sand from the roadside to cover the potholes, giving a false sense of repair. For their efforts, they stop motorists to demand tips. It usually takes one downpour to wash out all these surface measures. The state of the physical roads in Nigeria requires fundamental and structural solutions that can survive seasons and political leaderships. Similarly, the problems that may hinder a successful utilization of ICTs—all the technologies rather than just the cell phone—to achieve overall socioeconomic goals require structural solutions. Detours and surface measures may serve for a season, but the ICT4D project is slated for the long term.

Notes

[1] See Chapter 4 for a discussion of these policies.
[2] The scam takes its name from this: advance fee fraud.
[3] Thus it is common in Nigeria to hear a person or an action described as "419er" or "419."
[4] Tajudeen Diekola Oyawoye, personal interview in Abuja, November 2001 & May 2008.
[5] The World Bank, *World Development Indicators*, April 2002
[6] Ibid.
[7] Yesufu's analysis of the problems with Nigerian development efforts appeared in Chapter 3.
[8] In this chapter, the definition of culture is subsumed in that of ideology, defined here as a "set of ideas." While one acknowledges the differences between the two, one argues that they are not so starkly differentiated in the context of their connections with technology.
[9] As at the time of writing this (October 2008), I am still waiting for response to a query I sent to the NCC, NITDA, Federal Ministry of Education, and the Federal Ministry of Information and Communications through their websites four months ago.
[10] Ibid. The point about some IT journalists writing about what they do not know was observed in many articles that sought to "educate" the reader on e-mailing and the operations of Internet chat rooms. It was obvious that many journalists simply repeated expressions without really understanding them, even as one of the newspapers, *This Day* periodically published a glossary of ICT-related terms.
[11] The first is Nigerian Pidgin English for "I will drive myself" and the second means: "You can afford a car but can't afford to hire someone to drive you." In both cases, there is a societal assumption that an Oga—person of a higher socio-economic class—MUST have drivers rather than drive him/herself.
[12] Omoigui, Nosa, "The Information Age as a Key to Nigeria's Renaissance: Opportunities, Risks and Geopolitical Implications," Available at: http://www.nigerianscholars.africanqueen.com/opinion/NosaTech.htm
[13] "Niger Delta Development Commission: Historical Background," available at http://www.nddconline.org/history.shtml
[14] Omoigui, Nosa, "The Information Age as a Key to Nigeria's Renaissance: Opportunities, Risks and Geopolitical Implications," Available at: http://www.nigerianscholars.africanqueen.com/opinion/NosaTech.htm
[15] In many parts of the country, the *talakawas* (Hausa for the masses) in the North, "area boys" in the West and "Bakassi boys" in the East and "militants" in the South-South are paid by politicians (in military and civilian clothes) to foment violence as ways of achieving their political and often personal goals.
[16] Ekuwem, Emmanuel, in a personal interview in Lagos, November 2001
[17] Ndukwe, Ernest, in a personal interview in Abuja, October 2001.

• CHAPTER EIGHT •

'African Giant' Meets Africa: Information and Communication Technologies Beyond Borders

Introduction

So far, this book has addressed the development and diffusion of information and communication technologies (ICTs) in Nigeria. It has also examined the different approaches adopted by the country to harness these technologies for socioeconomic development. Although a latecomer into the global discursive arena on the utilization of the technologies for socioeconomic development, Nigeria quickly measured up, recording some of the highest growth rates in the diffusion and penetration levels of the technologies in Africa. The number of main telephone lines in the country grew at the cumulative annual growth rate of 12.8 percent between 2002 and 2007 (ITU, 2008). It was the ninth highest rate of the 54 countries in Africa and ahead of South Africa which actually recorded a negative growth rate of -1.7 percent during that five-year period. As Table 8-1 shows, Nigeria was not one of the top ten countries with the highest number of cellular phone subscribers per 100 inhabitants in 2007. However, it ranked eighth in the rate of growth in the cellular phone usage which constitutes 96 percent of total telephone lines, and 23rd in the percentage of the population who are mobile phone subscribers. The country's growth exceeded that of South Africa, 91.5 percent to 25.3 percent between 2002 and 2007.

Nigeria reached the ten million mark in the number of internet users, thus becoming the most internet-penetrated country in Africa by the end of 2008 (Internet World Stats, 2008). The diffusion of internet usage in Nigeria is spectacular considering an important factor besides its late entrance into the information society. The poor communication infrastructure in Africa generally has hindered "the deployment of broadband access via ADSL, which is the main method of fixed broadband access in most countries across the world" making the continent the least wired region in the world (ITU, 2008). It is therefore significant that Nigeria achieved the reported level of internet penetration within such a short time.

Table 8-1. Top Ten Mobile Cellular Subscribers in Africa

No.	Country	(000s)		CAGR (%)	Per 100 Inhabitants	As % of Total Telephone Subscribers
		2002	2007	2002-07	2007	2007
1	Seychelles	44.7	77.3	11.6	89.23	77.3
2	Gabon	279.3	1'169.0	33.2	87.86	97.8
3	South Africa	13"702.0	42'300.0	25.3	87.08	90.1
4	Algeria	450.2	2'762.7	127.7	81.41	90
5	Tunisia	574.3	7'842.4	68.7	75.94	86
6	Botswana	445	1'427.0	26.2	75.84	91.2
7	Mauritius	348.1	936	21.9	74.19	68.4
8	Libya	70	4'500.0	129.9	73.05	70.1
9	Morocco	6'98.7	20"029.3	26.4	64.15	89.3
10	Gambia	100	795.9	51.4	46.58	91.2
23	Nigeria	1'569.0	40'394.46	91.5	27.28	96.2

Source: International Telecommunications Union, 2008.

The relatively exponential growth of the ICT industry in Nigeria did not occur accidentally but it emerged through a deliberately unique policy framework. While several countries have formulated ICT policies especially following a 1996 African Information Society Initiative sponsored by the United Nations Economic Commission of Africa (UNECA), not many have recorded the degree of growth evidenced in Nigeria. The Nigerian policy structure is different from that of many countries on the continent in many respects. The Nigerian ICT policy approach is thus the focus of this chapter which discusses the uniqueness of the policy environment to highlight some "best practices" that might prove useful to other African countries as they seek to develop or finesse their own ICT sector. The chapter also examines particular areas where Nigeria is likely to take the lead in pushing the frontiers of the African ICT landscape. The underlying assumption is that Nigeria has much to contribute to a more robust ICT sector in Africa irrespective of the many potholes that litter its own pathways to the global information society and a prosperous future buoyed by ICTs. I begin the discussion with an overview of Nigeria's previous and ongoing role in Africa to provide a context for an examination of the prospects for future and continuing leadership in Africa. Past experiences of engagement can serve as a platform from which Nigeria can launch a twenty-first-century continental involvement as an active participant and catalyst in the socioeconomic development processes on the continent.

The African Giant and its Neighbors

Nigerians, with good reasons, like to refer to their country as the "Giant of Africa." With a population of 140 million, the country is the most populous on the continent. Nigerians constitute 25 percent of the Black race; statistically, one in four black people is a Nigerian (Elendu, 2006). With an adult literacy rate of 70.1 percent, Nigerians are among the most literate in Africa. Also, as noted in Chapter 3, Nigeria occupies a landmass of 923,800 square kilometres with as much as 90 percent of it being arable land. Combined with the varying climate in different regions, the arability of the land makes the country conducive for agricultural diversity and year-round cultivation of staple food and cash crops. Furthermore, Nigeria is rich in numerous mineral resources though the most economically viable, or at least the one that receives the most attention, is crude oil which generates about 85 percent of the country's foreign earnings. It also makes Nigeria the eighth largest exporter of crude oil in the world and a member of the Oil and Petroleum Exporting Countries (OPEC).

Paradoxically, it is Nigeria's oil wealth that frequently earns the country the sobriquet of the "Sleeping Giant" from other Africans. Its leadership has often been accused of corruption and mismanagement of resources such that despite its oil wealth, Nigeria remains on the World Bank's list of low-income economies—countries with $935 or less in Gross National Income. Its GNI per capita (Atlas Method) in 2006 was $790, lower than the average for sub-Saharan Africa, at $858 (World Bank, 2008). High prices of oil in 2008 affected Nigerians only in the skyrocketing inflation rate. A 50-kilogram bag of rice doubled in price in seven months in 2008; from ₦8,500 in January to ₦17,000 by August without a corresponding adjustment in earnings for average Nigerians. Nearly 90 percent of the population were living on less than $2 a day by the third quarter of 2008 (Ahemba, 2008). Nigeria has reportedly earned $1.2 trillion from oil production in 40 years, "the sort of money that enabled oil-producing Gulf states like Qatar to develop some of the strongest economies in the Arab world...Yet the vast majority of Nigeria's 140 million people live in no better conditions than their neighbors in West Africa" (Ahemba, 2008).

The perception that Nigeria mismanages its abundant resources fuels resentment in other Africans toward Nigerians. The sentiment intensified as the internal economic and political turmoil in the 1990s created a major "push" of migration of Nigerians to other African countries. This increased tension between Nigerian migrants and their hosts. Indeed, it was reported in 2008 that Ghana was threatening to deport Nigerians, perhaps to avenge a similar mass deportation of Ghanaians from Nigeria in 1983 (See, Aluko, 1985, for an excellent analysis of the "illegal aliens" episode, an interesting chapter in Nigeria's foreign policymaking). It could also be the case that with their population, generally higher level of education than other Africans and their "boisterousness," Nigerians have a reputation for "taking over" wherever they find themselves. This could partly explain the unique strain between them and black South Africans as the latter feel they have to compete for available jobs and services with Nigerians.

Other African countries have not been welcoming either and Nigerians continue to encounter hostility and outright violence abroad. In 2008 in Gabon, a Nigerian, Joseph Adumekwe, was roasted alive by security officials because he reportedly failed to produce another Nigerian wanted by the police for alleged theft. Fire was set on Adumekwe's back and he was later deported to Nigeria in the most inhumane condition (*The Guardian Newspaper*, 2008). There have been reports of other forms of police brutality and killings of Nigerians in African countries (Obadare, 2001). Even on a social level, conversations between Nigerians and other Africans in the Diaspora often quickly degenerate into expressed resentments over Nigeria's failure to utilize its resources in a manner that spills over to other African countries who are seemingly seeking an African Big Brother in the manner that the United States is to the world. To other Africans therefore, Nigeria is the "sleeping giant" of Africa and its role is considered irrelevant in the 21st century.

Nonetheless, the country has, in fact, acted as its brothers' keeper in specific important events and developments on the continent beginning with the nationalist struggles against colonization, political independence movements, and the creation of the defunct Organization of African Unity (OAU). In the latter process, Nigeria was part of the "Monrovia bloc," one of two groups that presented different approaches to the future of Africa during its decolonization in the late 1950s and early 1960s. The Monrovia bloc advocated a conservative, pro-Western ideology prioritizing economic and non-political union in the manner of the European Economic Community (EEC). The "Casablanca bloc," on the other hand, advocated a "political union or fusion of African states" (Ifidon, 2007). While there was a seeming compromise between the two groups, the principles of the Monrovia group arguably dominated the continental union that emerged in Addis Ababa in 1963. Those principles presumably accounted for the organization's seeming political weakness and ineffectiveness, particularly in maintaining peace in the most conflict-ridden continent in the world. After 39 years of struggling to be relevant in the turmoil that engulfed the continent during its lifetime, the OAU, previously amalgamated with the African Economic Community, in 2002, became the African Union (AU). The ascendance of the principles of the Monrovia bloc in the OAU was a credit to Nigeria's role in the establishment of the organization. The country therefore shared the blame for OAU's reputation as a "toothless bulldog."

Nigeria's achievement in the structure and function of the defunct OAU may be dubious, but its contributions to the struggle against the Apartheid regime in South Africa were unequivocally acclaimed. As Onishi (1999) noted, the oil money—considered both a curse and a blessing—enabled Nigeria in the 1970s and 1980s to become "the leading African opponent of white-ruled governments in South Africa, Namibia and the former Rhodesia, now Zimbabwe" (Onishi, 1999). Nigeria provided material support to the struggle and also allowed leaders of several African liberation movements and anti-Apartheid activists to operate within its borders.

In 1986, Nigeria led 32 countries to boycott the Commonwealth Games in Edinburgh, Scotland to protest the continued sports ties between the Margaret-Thatcher-headed British government and South Africa. The sports relationship between Britain and South Africa was only part of the broader call for comprehensive sanctions against Apartheid South Africa. Nigeria's anti-Apartheid efforts manifested through official state structures as well as the Nigerian civil society because Nigerian artistes (such as the late singer, Sunny Okosuns and his "Fire in Soweto" album), the media, individuals, students, trade unions and civil rights organizations joined the vanguard of the anti-Apartheid movement. As a member of the Eminent Persons Group, a Commonwealth mediation initiative, Nigeria's General Olusegun Obasanjo was a vocal anti-Apartheid activist in the 1980s and early 1990s. He cultivated friendship with jailed Mr. Nelson Mandela and visited him three times in 1986 in Pollsmoor prison outside Cape Town (Onishi, 1999). Also, Obasanjo used his diplomatic and political contacts (including friendship with American President Jimmy Carter) to pressure the international community to join the campaign against Apartheid in South Africa.

While waging war against Apartheid and colonization on the continent, Nigeria was also becoming involved in peacekeeping in the West African region. This was expanded through a military initiative under the auspices of the Economic Community of West African States (ECOWAS) that intervened in a bloody civil war in Liberia, 1989-996. The Liberian conflict, while unfortunate, succeeded in reinforcing Nigeria's role as its brothers' keeper and mediator of conflicts in the region. Despite the country's own "economic, political and social problems, (it) played a central role in forming the ECOWAS Monitoring Group (ECOMOG), a multinational military force which intervened in Liberia in 1990, Sierra Leone in 1997 and Guinea-Bissau in 1998"(*The Washington Times*, 1999). An international organization with no links to Nigeria also acknowledged, albeit reluctantly, the pivotal role that Nigeria played in achieving peace in Liberia.

> On 17 August 1996, after 134 days of killing and mayhem, Nigeria and other West African states brokered a ceasefire between the warring factions. Taylor emerged the dominant power, winning the 1997 presidential election. ECOMOG was dominated by Nigerian forces. General Sani Abacha, the corrupt ruler of Nigeria, enjoyed a good rapport with Taylor. Abacha persuaded Taylor to agree to the ceasefire and to participate in the election (Global Security Organization, 2008).

There is a general consensus that the ECOMOG forces, and especially Nigeria's role in providing the infrastructure and human and material resources, are fundamental to regional stability in West Africa. Nigeria's peacekeeping efforts have also extended beyond its regional sphere to other hot spots on the continent, such as Darfur, the war-torn region of Sudan.

It is therefore apparent that while Nigeria, starting with its promotion of the Monrovia Doctrine of non-interference and domination, was somewhat unwilling to flex its "gigantic" muscle, it has been thrown into situations where it has had no choice but to act as the Big Brother. This includes the anecdotal incident where in the 1970s, the federal government assisted a neighboring country by paying the salaries of its federal civil servants who had been owed months of unpaid remunerations. As a commentator noted, "Nigeria has committed itself to a number of pan-African processes...to put out the fires of conflict. It sees itself as an African leader...to their credit, they've shown some commitment to ending conflict and recognizing that there's a contagion effect for the whole of Africa" (Herbert, 2007).[1]

Nigeria has been influential in the continent on other fronts besides military. In West Africa, for instance, it is a major actor in both the formation of

ECOWAS and its operations mostly because it is the leading economy in the region. Though a lone English-speaking country in a sea of Francophones with a continuing distinct French influence, Nigeria remains the country that others want to be when they grow up. It was pivotal in the emergence of the Lagos Plan of Action (LPA), a short-lived Pan-African strategy of economic cooperation where African countries would be each other's major trading partners. The LPA fell apart with the debt crisis of the 1980s and the subsequent retreat of African countries to previously discredited externally oriented strategies of economic growth. These strategies remain the prevailing economic logic on the continent. Indeed, the most ambitious continental development plank in Africa, the New Partnership for Africa's Development (NEPAD), is anchored on strategic alliances with industrialized countries. NEPAD's principles include:

- African ownership and leadership;
- Anchoring the redevelopment of the continent on the resources and resourcefulness of the African people;
- Accelerating and deepening of regional and continental economic integration;
- Building the competitiveness of African countries and the continent. (Asika, 2002).

One principle that emphasizes "new partnership with the industrialized world" was underscored by the fact that NEPAD was on the top of the agenda of the 2001 summit of the Group of Eight (G-8) of industrialized countries and again in 2002 in Kananaskis, Canada. Three African heads of state, including Nigerian President Obasanjo, were invited to the summit and they left with promises of support from the industrialized countries in achieving the objectives of NEPAD.

> The main donor countries have...pledged to provide greater support to the continent—from aid and debt relief to trade and investment—so that the New Partnership for Africa's Development (NEPAD) can have a better chance of success than previous efforts. But as promise after promise seems to fade under closer scrutiny, Africans are still wondering what tangible support they will ultimately receive (Africa Renewal, n.d.)

As the African leaders returned to their countries aglow with the promise of new aid money and debt relief, the irony must have been lost on them that an initiative aimed at "anchoring the redevelopment of the continent on the resources and resourcefulness of the African people" depended for its success on financial resources from donors outside the continent. In the end though, the promised assistance did not arrive and where it did, it was only a trickle—to the consternation of many civil society groups (Accord International, 2008). Perhaps this is an indication that Africa's socioeconomic growth must truly be undertaken by Africans relying on Africa's human and material resources. This creates another opportunity for the "giant of Africa" to provide the initiative and leadership. To do that however, it must lead by example and first overcome its own challenges.

The country currently stands at a difficult crossroad: it does not offer a choice as both paths must be taken simultaneously. These paths manifest in the twin imperative for Nigeria to grow its internal economy and keep its population prosperous, on the one hand. On the other hand, it must be actively engaged in the economic and political affairs of the continent. How can Nigeria achieve these twin goals given the enormity of its internal challenges? This chapter suggests that the development and diffusion of information and communication technologies offer opportunities to accomplish this. The technologies possess the inherent capacities to accelerate national and regional socioeconomic development, as well as facilitate socioeconomic and political cooperation and integration on the continent. In the very least, ICTs facilitate communication—a necessary tool for continental cooperation and networking—and, if properly utilized, can potentially accelerate growth in key sectors. Already, Nigeria has taken the lead through the rapid expansion and diffusion of ICTs in the country. If the trajectory continues, perhaps it will become the continent's benevolent Big Brother and once again be the destination point rather than the origin of continental migration flows. This then raises a critical question: Are there any best practices that other countries can adopt from Nigeria's experiences so far in developing their own ICT sectors toward socioeconomic development? I lead into a discussion of the question with an analysis of the unique characteristics of Nigeria's ICT landscape.

Key Factors in Nigeria's ICT Growth

Specific policies and enabling laws—the NITDA Act and Telecom Act—were the twin engine that propelled developments in the telecommunication and general ICT sectors. The nature of the drivers was equally crucial. During the debates leading to the formulation of the governing policies and law, stakeholders argued strongly for the involvement of the private sector in the industry. Many commentators in the media repeatedly argued that "government had no business providing phone services." They cited the problems with public utility services in Nigeria which ineptitude had always been blamed on the direct involvement of the government. Currently the power generation and distribution, and production and supply of petroleum products in the country suffer from this direct involvement of the government. In the view of principal actor in the country, Titi Omo-Ettu, a telecommunications engineer:

> It is wrong for the government to run (public utilities). For instance, to use ICTs for development, the government should provide enabling environment for private sector to thrive through regulation, fairness, protect investments, and (ensure that) laws are obeyed and complied with. Government should try to help the service providers by providing market for them by patronizing these businesses.

A private-sector driven approach created the environment, initially for the mobile telephone industry to take off and flourish the way it did. The NITDA Act of 2007 also gave the Nigerian Information Technology Development Agency (NITDA), the agency implementing the IT policy, the mandate to expand its activities from its headquarters in Abuja to the 36 states. These activities included creating awareness of the advantages of ICTs, building capacity through various trainings and workshops and developing a public service information network (PSnet). The network involved the deployment of:

- A VSAT hub at NITDA's head office;
- Wireless hot spot in Abuja, including Wi-Fi in the federal secretariat building;
- Interconnection of key federal government buildings with fibre optic cables;
- VSAT remote terminals in 36 states (15 states were already completed by 2008);
- Broadband wireless extensions in 27 states;
- Routing and switching facilities at NITDA's Abuja office and in the 36 states;
- Development of collaboration applications such as video conferencing, messaging, distance learning and security.

The Nigerian Communications Commission (NCC) was most instrumental in the growth of the industry because of its overarching portfolio particularly in licensing and general regulation. It supervised the auctioning and issuing of the initial four digital mobile licenses in 2000–2001. By 2006, after a mandatory five-year protectionist period, the Commission called for bids for unified licenses which allowed all operators (who met certain criteria) to provide both mobile, fixed wireless and data services. This increased the number of providers to 12 including two national carriers. During the first five years, the digital mobile phone operators were banned from sharing infrastructure and each company was encouraged to develop separate infrastructure. They were also not allowed to form an association. These restrictions were aimed at discouraging the operators from predatory practices such as "ganging up" against the government and consumers. The providers were also required to expand into the rural sector to avoid high concentration in big cities where operational costs were lower. While a significant percentage of the country does not have access to telecommunications services, the operators are present in all 36 states and Abuja, therefore fulfilling their geographical spread requirement.

The NCC also issued 26 licenses including those for the digital mobile, fixed wireless access and unified access service. Others are: internet service provision (ISP), cyber cafés, sales and installation of terminal equipment, community telephony, payphone operation, international gateway, public mobile communications trunk radio services and VSAT (hubs, domestic and international). The Commission strictly supervises the operation of these various licenses in the same manner that it does the telephony sector. It turned out that while the formulators of the relevant polices and legal frameworks had conceded to creating private-sector friendly ICT environment, the NCC kept operators under a tight rein, even as consumers criticized the regulatory body for being too pro-operators to be as transparent and fair as it claimed. The organization responded to the criticisms by creating a telecom consumer parliament as well as a consumer affairs bureau (CAB) thus also positioning itself as a watchdog for consumers. The bureau provides information for consumers while also processing complaints against service providers. Indeed, on its website, the Commission has a database of complaints and actions taken. It also publishes names, addresses and phone numbers of consumers in a manner that indicates that Nigerians, still euphoric about the new technologies of communication and information, are yet to consider privacy issues.[2] It then becomes ironic that while the Commission seeks to address consumers' problems, it unwittingly exposes them to bigger problems when their personal information is made so easily available to the internet public. In any case, NCC's unique combination of deregulation and control was a critical factor in the rapid diffusion of ICTs in Nigeria between 2002 and 2007.

Another explanation for the remarkable expansion of Nigeria's ICT sector is that the services were long overdue. In other countries, such as South Africa, there was relatively high teledensity and therefore people used the telephone, for instance, more generally than had occurred in Nigeria. As noted in Chapter 6, in August 2001 before the digital phone operators rolled out their services, there were less than 500,000 active phone lines in Nigeria, and fewer than 100,000 mobile phone lines. The process and cost of obtaining a phone line—fixed or mobile—were so complex and costly that only the very rich could afford one. By the end of 2001, there were one million cell phone lines in the country, exceeding the most optimistic estimates only four months earlier. As Nigerians patronized a much-needed service, the cost began to drop such that by the end of 2008, one could obtain a cell phone and line for less than ₦4,000. While this was still expensive relative to the income of the average Nigerian, owning a cell phone had also become a basic tool of communication for many as well as a status symbol. Indeed, the question was no longer whether one had a cell phone, but how expensive, how sophisticated and how many handsets.

Furthermore, a major problem in the telephony sector paradoxically generated increased subscriber base. While the infrastructure developed and improved rapidly during the crucial 2002-2007 period, the sector was prone with poor interconnectivity between the networks and unreliable access outside major cities and urban centers. To overcome these problems, users began to acquire multiple cell phones and subscriptions, both to take advantage of cheaper in-network calling and to ensure successful connection. Thus when a user wants to initiate a call, the destination number determines what phone to use. The networks have identifying prefixes making it easy for users to recognize the different networks from the phone numbers. It is therefore common to see Nigerians with four different cell phones. Actually, it was unusual for people of certain socioeconomic status to have only one cell phone as a cab driver told me in January 2007 during the second phase of my field research. In a conversation with him on the different cell phone providers, I mentioned that I was with Celtel, one of the providers, now known as Zain. He asked, "And?" Puzzled, I wanted to know what he meant.

"I mean, you have Celtel and MTN or Glo?" When I told him I had only Celtel, he turned around to stare at me. It was now his turn to look puzzled.

"What do you mean...that you have *only* Celtel?!"

Multiple mobile phone subscription is common in Nigeria and other African countries. It solves some problems but creates another: inability to accurately determine the number of mobile phone lines on the continent. By the end of 2007, there were 260 million cellular subscribers in Africa (85 percent of the phone lines on the continent). This made Africa the region with the highest mobile cellular growth, at an annual growth rate of 65 percent (ITU, World Telecommunication/ICT Indicators, 2007). The numbers were expected to reach 300 million by the end of 2008.

> However, some of those customers may have been counted more than once and some may be neither citizens nor even residents of the countries in which they are counted as customers...The question policy makers have to address is how to interpret such

very large numbers and how they reflect the reality of their countries, cities, towns and villages. In particular they must consider whether they are achieving the Millennium Development Goals (MDGs) for which mobile subscribers per 100 population is a target indicator" (Sutherland, 2008: 1).

The phone companies do not or cannot "unduplicate" the multiple subscription numbers because the large numbers give them competitive age over each other. Also, as Sutherland notes, the large numbers bolster the image of phone companies especially when they are making a case to the government (2008). Similar to the issue of privacy in an ICT world, multiple subscriptions do not currently present any major concern for ICT actors especially because there is also multiple-usage as epitomized by the umbrella people phenomenon. In large cities such as Lagos, it may seem that everyone has at least one cell phone. But obviously even with 50 million cell phone subscribers in Nigeria, about 90 million people are still left without direct access. As with most problems in Nigeria, there is always a coping mechanism; in this case, it is the "umbrella people" who step in to plug the gap. With more people having access to a telephone through these men and women, it would seem that in reality there are more users of the technology than the official numbers might indicate, even when multiple subscriptions are factored in.

The umbrella people also perform interesting service beyond phone calls or repairs by acting as money transfer agents. Usually a person seeking to remit money to someone else in another part of the country buys a phone recharge card, texts the number to the recipient who then sells it to an umbrella person. The umbrella person confirms that the number is valid and then pays the cash equivalent to the person, after deducting the service charge. The transaction is cheaper and faster than if the customer utilized official money transfer agencies or went through the banks. This service is mostly patronized by people who do not have bank accounts or identification cards and are only transferring small amounts of money. This form of phone-enabled financial transaction has been perfected by Safaricom Kenya, operators of the increasingly popular M-Pesa in Kenya. It is a mobile banking service that allows customers to remit money at a service fee of about one dollar. The company expanded so rapidly that after one year of operation it sought to open its services in Britain, from where much of its remittances originate (Aron, 2008). While the more formalized approach raises the status of this unique mobile banking sector, the Nigerian street-level equivalent meets the basic needs of the average user and creates the potential for employment for many people.

Given the variety of everyday services that the umbrella people provide, other African countries can facilitate faster diffusion of ICT usage through the establishment of a framework conducive to their activities. They can choose the option of the Nigerian informal variety, the Kenyan higher-scale method or the many variants that can be generated from the two major approaches. Of course, given the overwhelming presence of the informal economy in many African countries, the technologies and their applications are likely to take off on their own as users and consumers shape usage to achieve maximum utility. However, there is also benefit in structuring the appropriate environment to facilitate innovations of usage and sophistication in the applications. This is where the Nigerian experience may help as other countries on the continent develop their own ICT industries.

In summary, three major factors have shaped ICT development in Nigeria and can potentially appeal to other countries on the continent. These are: active participation of the private sector, a policy and legal framework that deregulates but still retain operational control of the industry, and the eagerness of Nigerians to adopt technologies that were long overdue in the country.

Prospects and Challenges for Continental Leadership

In an informational mode of production with the emphasis on knowledge and service economy, Nigeria has an advantage, theoretically. It has the population, relatively high literacy levels and material resources and innovativeness. As we have already noted in this chapter, Nigeria has come from behind to emerge a major actor in the ICT industry in Africa. At the policy level, it has developed a unique approach of quasi deregulation and control: it has allowed private sector interests to enter the market and freely compete; at the same time, regulatory agencies have a fairly tight rein on the process, freely sanctioning service providers when they get out of control. This is a different approach from the policies adopted by other some other countries. In South Africa, the government holds so tightly that the system is almost ineffective, and may account for the slow and negative growth rates in some sectors of the ICT industry (*Business Day*, 2008). Ghana seems to waver between a public-sector controlled and fully deregulated ICT industry. This manifested in government's attempt to build a national fiber optic backbone and operate it as a public utility. And then it received a $30-million "gift" from the Chinese government to fund it. Chinese companies are aggressively investing—some might say, taking over—in key African ICT markets. One of these companies is Huawei Technologies which has its tentacles all over the continent. It sees "Africa's economical and social factors, such as its young populations and the low mobile and internet penetration" as providing "many opportunities to address the enormous potential and respond to the sophisticated demands by the local operators and consumers" (Feng, 2008). A cynic might therefore argue that a "gift" from the Chinese government to build telecommunication infrastructure in an African country is a calculated investment in a viable market.

Nigeria's approach seems to have succeeded and this accomplishment could position it to lead the vanguard of ICT development and utilization of the technologies for socioeconomic development in Africa, just as the country spearheaded a continental anti-Apartheid movement in the 1980s and 1990s. So many foreign companies are moving into Africa to exploit a huge market with customers eager for long awaited technologies. While companies like Huawei suggest that their investments in Africa are mutually beneficial and make a commitment to Africa and "to supporting the local economies of every country we operate in," ultimately they are not in the business for charity. Investments by Africans or African companies in Africa have a greater chance of benefiting the majority of the local people. There is therefore a leadership vacuum in Africa into which Nigeria can step.

This presupposes the capacity of ICTs to facilitate Nigeria's capability to act as Big Brother in the West African region and beyond. For one, an ICT-enabled economy would have enormous impact on the surrounding economies of West Africa with many of the countries already depending on Nigerian products. Indeed as argued by Akpan-Obong and Parmentier (2008), ICTs promote the achievement of both national and regional economic goals. Many elements of ICTs particularly foster closer regional integration (which in turn creates economic growth) through ease of communication and cross-border processing of movements of goods, services and labor. The specific goal of integration and cooperation in the various regions of the world is economic growth by itself as well as facilitation of competition in the global economy. Information and communication technologies potentially enhance integration as a key to economic growth as well as act as tools for economic growth directly. The possibilities are universal but even more so in Africa where regionalism might very well be its major coping mechanism in an increasingly globalized and "flat world" where only the fast, biggest, strongest and fittest will survive (Friedman, 2005). As its ICT policies emphasize, Nigeria seeks to have a competitive edge in the emerging global network society. The path to achieving this goal is by making itself the "most able to innovate, compete and win in the age of globalization," as Friedman (2008) writes in the context of the American economy.

There are several pathways that Nigeria can take to spearhead a continental strategy to give Africa a competitive edge, not only in the global economy, but in its ability to feed its populations and keep the peace on the continent.

After all, most of the conflicts on the continent are over resources—or the lack of them. An Africa self-sufficient in its basic needs will be a more peaceful continent. Nigeria may very well be the only country on the continent with the material and human resources to turn Africa into an ICT continental enclave.

From the field research, it is evident that much of the application of the technologies in Nigeria are still at the basic or conventional level of usage though more sophisticated adoption and adaptation were also observed. Numerous small-scale businesses have sprung up around ICTs either in sales or services. Some manufacturing has occurred with the presence of companies such as Zinox Technologies, though a significant percentage of ICTs used in the country are imported from China and the United States. The Hewlett Packards and Microsofts of the global ICT production market have set up shops in the country, and their representatives and lobbyists parade the corridors of government and power seeking supply, consultancy, service and training contracts. For this reason, the country is clearly not yet an ICT giant even if its consumption of ICT products and services are gargantuan in proportions. For the country to take the lead in ICT development on the continent, production of the technologies and innovativeness in usage must be a top priority on the agenda of policymakers, regulators and other stakeholders.

At the risk of repetitiveness, I emphasize the point that Nigeria has the resources to accomplish this; it just needs the political will to diversify into this area. The Nigerian movie industry, known as Nollywood, is evident of what Nigerians can do when they are determined. The remarkable thing about the Nollywood phenomenon is that it developed outside the ambit of the government. It started from small home-video-type technologies and basic scripts and poor acting in the 1990s to become the third biggest movie industry in the world (behind Hollywood and Bollywood, the Indian movie industry) in less than 20 years. The production, storylines and acting have grown progressively over the years such that the industry has caught the attention of Hollywood actors. Nollywood products are now marketed all over the world, especially in other African countries. Production of ICTs can follow Nollywood's trajectory—begin with basic repairs, assembling, and basic technologies and end up with high-end technologies that will not only make Nigeria ICT-self-sufficient but enable it to be an exporter of the products to other countries on the continent. These activities can also enable Nigeria to be Africa's Silicon Valley and an ICT Mecca for other Africans. In the process, the country can present a much-needed leadership to other countries in the utilization of ICTs to promote national and regional socioeconomic development as well as regional and continental integration and cooperation.

Currently there are several initiatives aimed at encouraging the manufacturing of ICTs—both the hardware and software. While most hardware manufacturing is assembling, there have been other efforts at actual manufacturing. For instance in 2001, Zinox launched the "first Made-in-Nigeria" computer. In the years since then there have emerged several "computer villages" and the Abuja Technology Village, Nigeria's equivalent of the Silicon Valley. The federal government encourages these endeavors by requiring its agencies to purchase only locally produced ICT hardware and software. Where required technology is not locally produced, government officials are expected to purchase only from companies with a high number of Nigerians in management positions. With these initiatives—and others underway—Nigeria can position itself as a major producer and exporter of ICTs to the rest of the continent. That is, as soon as it can wean itself from relatively cheaper imports from China, India and the United States. Until then, the computer villages are likely to remain sites where the only "productive" activities that occur are in repairs, servicing and assembling of foreign made products. These villages presently serve as centers for the installation of software and upgraded hardware into used computers either imported from abroad or "donated" to Nigerians. The influx of used computers in the Nigerian market presents its own challenges, reminiscent of the 1980s practice of dumping nuclear wastes in Nigerian coastal towns by Italian companies. Most of the used computers currently being "donated" to Nigeria by individuals and organizations in western countries are the environmental disaster of the near future.

A second area where Nigeria can take the lead is in the development of content on the internet, a technology where its usage has overtaken that of other African countries. As at June 2008, according to the World Internet Statistics, Nigeria had the most internet users in Africa (Figure 8-1). Former President Obasanjo in the early days of his "internet innocence" had joked about how everyone was talking about downloading from the internet. He asked at a press conference in 1999: "When do we upload?" The audience laughed and many derided the president for being naïve about the internet. In some ways, Obasanjo did ask a serious question—one that has been asked more frequently in recent years. African countries, as in everything else, continue to be the consumers of ICTs, or in this case, the downloaders of information on the internet. It has uploaded (contributed) an insignificant amount of content to the internet. This has raised unease in some quarters that Africa is simply standing on the sidelines (as it did in pre-colonial years) while the rest of the world stakes out their frontier on the internet. Muchie (2008) suggested that we may be entering the new age of the scramble for Africa because the continent is again becoming a passive participant in the events that will be pivotal to its future survival. "Everyone has an initiative or something *for* Africa but Africa has nothing for itself," he said at a conference on "Confronting the Challenge of Technology for Development: Experiences from the BRICS" in Oxford, England in May, 2008.[3] Much of the discussion on Africa and the internet have centered on access and affordability. There is a certain conventional wisdom that suggests that there is a world of information out there on the internet that would be useful for Africans—if only they had the access. Many ICT advocates wax lyrical when they speak about how information on planting season can help the local African farmer—if only she had access to the internet. For sure, such euphoric claims have receded over the years and there is more critical examination of the question of content. For instance, a 2006 research by the African Internet Service Providers Association (AfrISPA) showed that Africa is lagging behind in "the creation of useful and relevant local information for Africans living and working in Africa" (van Reijswoud, 2006). The Association suggested some strategies for promoting African content on the internet. These included:

1. The development of sustainable content models that respond to what the consumer needs and wants...
2. A regulatory and policy framework that supports content creation, (governmental) content digitization, content sharing and free speech, and protects the authors.
3. Implementation of tools, like national portals, that enhance the visibility and improve the search-ability of local content.
4. Lower barriers for access of content for non-English speakers. (AfrISPA, 2006)

The first strategy will necessarily require a huge ideological leap from the prevailing practice in many African countries where the leadership decides the needs of the people. Already embedded in the ICT4D discourse are some of the top-down strategies of growth that dominated classical development theories. As noted in a previous chapter, the information society, by definition, breaks down hierarchies such that there is no separation between authors and readers/audience, kings and subjects, masters and servants, political leaders and citizens, teachers and students. This is underscored by the explosion of Wikipedia and blogosphere where everyone is a content provider. Yet, many Africans continue to download from the internet while contributing a negligible percentage of the content. Nigerians and government ministries and agencies have websites rich with active content—though many of the government websites are long overdue for updating. Further advances in creating local (and interactive) content on the internet would be useful contributions to the diffusion of internet in the country, not just for its sake, but as a tool for national and regional socioeconomic development, and continental cooperation and integration.

Figure 8-1. African Top 10 Internet Countries

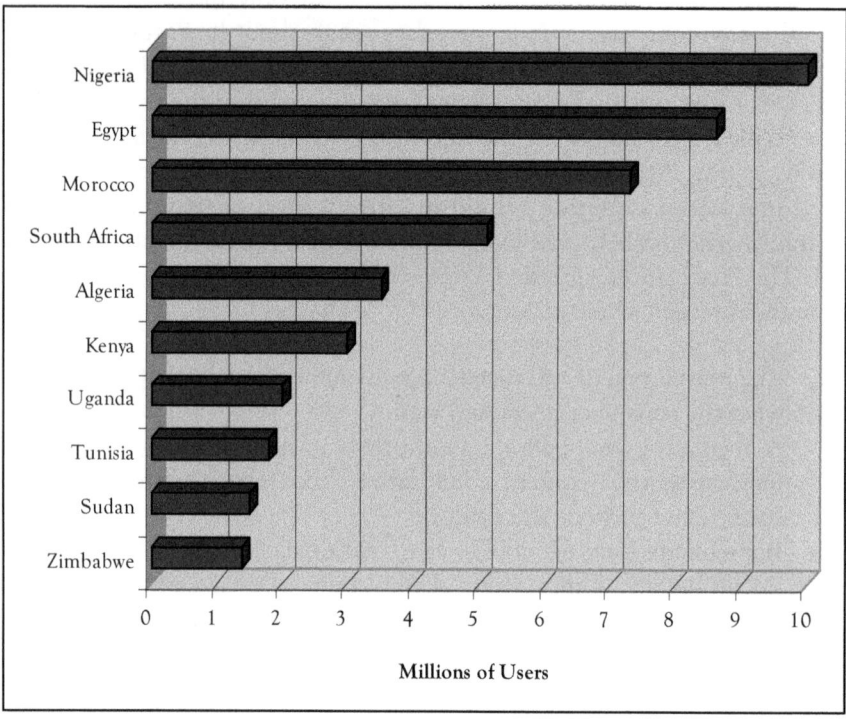

Source: Internet World Stats, 2008

Nigeria can play a critical role in developing relevant local content on the internet as well as build infrastructure to facilitate wider access not only in the country but on the continent as well. As the *Economist* notes, besides the "stultifying lack of local content," Africans "must also cope with obsolete systems, irregular ... Interfaces are being written in a number of African languages, but even the clearest instructions in Wolof or Yoruba as to how to use Windows presume a fair degree of literacy" (2007). The International Telecommunications Union was so concerned about the status of Africa on the internet that it called for a "Marshall Plan" in 2007 as a strategy not only to achieve a higher penetration level but also to reduce gaps in infrastructure that make internet access in Africa the most expensive in the world. As I noted earlier, 70 percent of African internet traffic passes through networks in Europe and North Africa (Forbes, 2007). This is likely to change soon as various countries on the continent begin to build domestic internet exchange points (IXP). Already, through the SAT 3 undersea cable on the West Africa coast, Nigerian internet users enjoy faster internet speeds than in other countries. An IXP allows various ISPs to exchange internet traffic between their networks by means of joint peering agreements, which enables messages to be exchanged without additional expenses (UiXIP, 2008). It enables networks to interconnect directly, via the exchange, rather than through other networks. In the absence of an IXP, connections between ISPs in a country are routed through another country to a global connection thus increasing transaction costs and time. This partly explains why internet access is so expensive in Africa. With an IXP, bypassing a third party reduces the demand on a country's international bandwidth thus reserving it for actual international traffic. Several Sub-Saharan African countries, including Nigeria, have now developed IXPs to manage local traffic.

For any project at creating local content on the internet to succeed, Nigeria's major asset—the people—must be active participants in the process. However, they may also be a major liability. As noted earlier in this chapter, Nigerians, as a group, are not the most loved people in Africa. It would be easy to dismiss the resentment that other Africans have toward Nigerians as envy but some introspection shows that there is more to that. While the majority of Nigerians are hard-working and resilient people, a few have gotten involved in practices that have brought a certain infamy to Nigerians: fraudsters. Before the internet arrived in Nigeria, the country was notorious for the "advance fee fraud," a scam where someone sends a letter to a foreigner and informs the recipient that there is unclaimed money in Nigeria—usually in millions of dollars—that the foreigner can have simply by paying the processing costs and providing his or bank information. The usual storyline is that someone who was either deceased or could not be located was owed money for contract he or she had done for the federal government years earlier. All that the foreigner needed to do therefore was claim entitlement to the unclaimed money. The scam artist would promise to provide all the documents the foreigner needed to authenticate his or her claims. Countless foreigners have fallen victims to the crime known in Nigeria as "419" because they themselves were greedy enough to seek to reap where they did not sow. In the past, the solicitations were made by posted letters and faxes. These days, it is through e-mail. The fraud has spread such that law enforcement agencies in different countries have established units just to investigate Nigerian 419 activities. Travel alerts on countries' websites warn their citizens against doing business with Nigerians.

A Nigerian in the Diaspora wrote in a post to a Nigerian news forum that the Nigerians are now synonymous with fraud as cyber cafés in Nigeria have become "dens for internet fraudsters" who utilize their internet access to scam people around the world.

> I receive more than thirty of these scam letters everyday and several American colleagues have also reported doing so. The world over, none in his or her right mind wants to do any transaction with anybody who is from Nigeria...The scam originating from Nigeria is more than an economic problem it is a national security issue and something must be done (Ette, 2008).

He noted that while the Lagos State police department has a task force dedicated to the activities of internet scam artists, a more functional solution was required to get rid of the "national embarrassment." Rather than a solution, the phenomenon has actually taken a different, more insidious turn in recent times.

Fraudsters from Nigeria, and increasingly Ghana, now go to online dating sites to snare people looking for love, marriage or just friendship.

> It's one thing to get scammed for responding to an e-mail from someone promising gazillions of unearned money. But the need for love is very human and no one should be scammed for it... [T]hese scam artists register and create their profiles, post photos (of other people) and lie about their location and occupation. Men pose as women to snare men, and Nigerian and Ghanaian 419ers pose as white men to attract white women. Their profiles are matched with genuine seekers and (and probably with other scam artists too). (Akpan-Obong, 2008)

The Queensland Police Service reported the arrest of a 23-year-old Nigerian who posed as a woman at an online dating site and swindled an Australian man of $20,000. The report said the Australian law enforcement agents worked with the Nigerian Economic Financial Crimes Commission (EFCC) to ferret out the Nigerian. The Queensland minister of police, Ms. Judy Spence, said her police force is "leading the world when it comes to the investigation of Advance Fee Fraud internet-based offences" (Stuff, 2008). She then cautioned the public against responding to "Requests to send money or personal information like account details overseas to an online companion"(Ibid.) When the fraudulent love interests are not demanding outright money, they ask their preys to receive goods purchased over the internet and re-ship the packages to them, as a CNBC news program revealed in September 2008. Since some courier companies will not ship to Nigeria, the fraudsters befriend people—mostly lonely single males—in different parts of the world. They purchase small, but very expensive, items online and use the shipping addresses of their overseas "friends." After the items have arrived, they request that they should be re-shipped to them in a neighboring African country where the fraudsters then take delivery. Most of the items are purchased over the internet with stolen credit cards.

These fraudulent activities are undertaken by an insignificant percentage of the Nigerian population yet the whole country gets tarred with the same brush. This generates major obstacles for genuine businesses whose activities may require collaborations and strategic partnerships with international businesses or agencies. Trust is an important factor in business, especially in the virtual economy where individuals may not have the benefit of personal contact to establish trust. In an industrial-age-based economy, a man could never pass for a woman to swindle $20,000 from another man who believes he is helping a female love interest out of financial difficulties.

It is not clear how much the activities of scam artists have cost Nigeria in monetary terms. However, it is indisputable that they have undermined the confidence of foreign investors in the Nigerian business environment thus creating problems for genuine entrepreneurs in the country. This compounds the already trust-deficit business environment within Nigeria itself. While credit card and ATMs have been introduced in the country and banks offer internet transactions, the economy is still essentially cash-based. Cash seems to be the only currency of value and trust. Even in this scenario, people still exercise vigilance because there have been instances where bank tellers short-change customers who may not have the patience or presence of mind to count huge amounts of cash before leaving the bank.

The EFCC has a cyber crimes unit to monitor the activities of internet scam artists, also known as "Yahoo boys" (probably because they conduct their activities through their Yahoo e-mail accounts). The major strategy adopted by the EFCC to reduce the activities of the "Yahoo boys" was to ban overnight cyber cafés and enforce a standard closing time for the businesses. Also, cyber café owners were co-opted into policing the activities of their customers. The policy shifted the burden of policing to business owners whose interests are best served by creating a welcoming environment for their customers. While Nigerians are not obsessed with privacy issues, it would be disconcerting for customers if the cyber café owner is standing over their shoulders reading their e-mails. The mandatory closing time has therefore not been rigidly enforced by cyber café proprietors as that would certainly force customers to go elsewhere. Also, the rule does not discriminate between "Yahoo boys" and other patrons of cyber cafés who may be conducting genuine business at times most conducive to them. The EFCC did create a cyber crimes unit, but expectedly, much of the commission's resources are directed at relatively bigger issues such as investigating and trying former governors accused of corrupt enrichment while in office. In the absence of a functional set of mechanisms to deter the activities of the "Yahoo boys" and restore integrity to the Nigerian business environment, genuine Nigerian businesses must earn the trust of foreign investors on their own through a variety of approaches such as establishing networks around the commercial attachés of the different embassies and high commissions in Nigeria; international chambers of commerce, alumni associations of foreign universities (for those who studied abroad) and strategic alliances with reputable Nigerians in the Diaspora. Certainly, more needs to be done in this area by the Nigerian government to repair the country's international image and reputation especially with other African countries. It is only by doing so that the achievements that Nigeria has recorded in the ICT sector will be worth emulating.

At this juncture in the third quarter of 2008, it does not seem like there is anything to be learnt from Nigeria. For one thing, it is a relatively new entrant into the ICT industry though its growth has far exceeded those of many other countries on the continent. However, given Nigeria's history of pan-Africanism starting from the decolonization movement, the creation of the OAU, fight against apartheid and peacekeeping, the country can once again rise to the top and be the giant of Africa that its citizens claim it is. It has the human and material resources to achieve this. The contention in this book has been that ICTs are not the panacea to any country's development problems as the ICT4D discourse might indicate. Indeed the issues of development that confront Nigeria, as well as the rest of Africa, are so complex that a multi-pronged approach is required to achieve any sustainable socioeconomic growth. Still, the technologies present real possibilities for Nigeria to achieve its own socioeconomic development and foster regional and continental cooperation and integration, as a dual approach to providing leadership once again in Africa.

Conclusion

Even before Nigeria achieved political independence in 1960, its nationals were already involved in various pan-African movements to end colonization and return Africa to Africans. For instance, Dr. Nnamdi Azikiwe, a foremost nationalist was nicknamed "Zik of Africa" because of his pan-African crusade to end African colonialism. The anti-colonization struggles by Nigerian nationalists transcended their national borders. After independence, Nigeria, as a country, and many of its citizens continued to be actively involved in the broader issues of concern to Africa. Due to the country's own internal political, economic and military crises, Nigeria was focussed inward and was not always consistent in playing the role of the "Big Brother" such that the Giant of Africa was re-named the "Sleeping Giant" by other Africans. Advances in ICTs offer Nigeria the opportunity to re-introduce itself to the rest of Africa. To accomplish this however, the sleeping giant has to wake up. The following wake-up calls may be required to make Nigeria once again relevant in the affairs of Africa: be a producer of ICTs; build ICT infrastructure; create effective mechanisms to counter the negative activities of the "Yahoo boys" so that the internet is not synonymous with scam; sanitize Nigeria's image in the international community and create trust in Nigerian businesses; develop a robust national economy that meets the basic needs of the majority of Nigerians. Then, the rest of Africa will behold once again the Giant of Africa.

This book is a compendium of the development of ICTs in Nigeria both as means to socioeconomic development and as ends by themselves. It is a noteworthy project if only to showcase where Nigeria began and how it got to where it is today. As a US politician said in 2008, we cannot ignore the past because it is the prologue to the future. In that sense, this book is the prologue of current utilization of ICTs for socioeconomic development in Nigeria and the future when subsequent research can evaluate the outcome of its premises. A documentation of the development and accomplishments of the Nigerian ICT sector is a worthwhile by itself; however, it must go further and provide some practical lessons for other African countries. I conclude this chapter—and the book—by arguing that Nigeria has indeed developed, deliberately or accidentally, an interesting policy approach to growing its ICT sector.

Notes

[1] Ross Herbert, an analyst with the Johannesburg-based South African Institute of International Affairs was speaking to the Associated Press's Edward Harris, October 6, 2007.

[2] The lack of consciousness of privacy is such that letters to the editor in newspapers not only carry the writers' home addresses but phone numbers as well. Also, the personal pages of newspapers publish phone numbers of those writing in to seek spouses or other love interests.

[3] Mammo Muchie (Aalborg University, NRF/DST SARCHI, Tshwane University, Pretoria) expressed this view while speaking during a discussion at one of the conference's plenary sessions.

Bibliography

Adebo, Simeon (1966). "Foreword" in Stolper, Wolfgang F., *Planning Without Facts: Lessons in Resource Allocation From Nigeria's Development*. Cambridge, Massachusetts: Harvard University Press.

African Internet Service Providers Association (2006). "Strategies for increased internet growth: A call for a paradigm shift to stimulate internet growth through content." Retrieved Sept. 27, 2008 from http://www.regulateonline.org/content/view/861/76/

Africa Policy Information Centre (1996) "Africa on the Internet: Starting Points for Policy Information." Retrieved on July 4, 2000 from http://www.africapolicy.org/bp/inet.html

Africa Renewal (n.d.). "Funding for NEPAD: Africa still waiting for genuine 'partnership.'" Retrieved Sept. 24, 2008 from http://www.un.org/ecosocdev/geninfo/afrec/sgreport/partner.htm

Ahemba, Tume (2008). "Squandered oil wealth leaves Nigeria in the dark age." Reuters, July 21, 2008.

Aka, Ebenezer, Jr. (1999). *Regional Disparities in Nigeria's Development: Lessons and Challenges for the 21st Century*. Lanham, Maryland: University Press of America.

Akpan-Obong, Patience (2008). "Love, the new currency of scam." *Saturday Punch*, July 12, 2008, back page.

—— (2009). "Unintended Outcomes in Information and Communication Technology Adoption: A Micro-level Analysis of Usage in Context." Forthcoming.

—— and Mary Jane Parmentier (2008). "Coordination, convergence or contradictions: Information and communications technologies for integration and development in Southern Africa and the Southern Cone." Paper presented at the Sanjaya Lall Programme Conference on Confronting the Challenges of Technology for Development: Experience from the BRICS, held May 29-31 at University of Oxford, England.

Almond, Gabriel and Sidney Verba (1965). *The Civic Culture; Political Attitudes and Democracy in Five Nations, An Analytic Study*. Boston, Massachusetts: Little, Brown.

Aluko, Olajide (1985). "The expulsion of illegal aliens from Nigeria: A study in Nigeria's decision making." *African Affairs*, Vol. 84, No. 337 (Oct., 1985), pp. 539-560.

Amaefule, Everest (2001). "Zinox, a toast to private sector initiative," *Daily Times*, Thursday, October 18, 2001, p. 15.

Anunobi, Fredoline O. (1992). *The Implications of Conditionality: The International Monetary Fund and Africa*. Lanham, Maryland: University Press of America, Inc.

Aragba-Akpore, Sonny (2000). "NITEL, MTS may be favored for mobile phone permits" in *The Guardian Online* Previously available at http://www.ngrguardiannews.com

—— (2000). "NCC has powers to issue all telecom licences, says Arzika," *The Guardian*, Monday, July 3, 2000, pp. 58-59.

Aremu, 2008 from ch. 3 (confirm)

Aron, Morris (2008). "Kenya: M-Pesa's bid to enter UK runs into legal hurdles" in *Business Daily*, (Nairobi). Retrieved Sept. 29, 2008 from http://allafrica.com/stories/200803240968.html

Asika, Chinyere (2002). "What is NEPAD?" Retrieved Sept. 23, 2008 from http://www.nepad.org.ng/PDF/speeches/brief.pdf

Bada, Abiodun (2002). "Local Adaptations to Global Trends: A Study of an IT-Based Organizational Change Program in a Nigerian Bank," in *The Information Society*, Volume 18, Issue Number 2.

Balsamo, Anne (1996) "Myths of Information: The Cultural Impact of New Information Technologies," *Technology Analysis and Strategic Management*, Volume 8, Number 3.

Bates, Robert H. (1988). "Governments and Agricultural Markets in Africa," excerpts from Toward a Political Economy of Development reprinted in Mitchell A. Seligson and John T Passé-Smith, eds. (1998) *Development and Underdevelopment: The Political Economy of Global Inequality*, 2nd edition. London, UK: Lynne Rienner Publishers, Inc.

Bell, Daniel (1973). *The Coming of the Postindustrial Society: A Venture in Social Forecasting*. New York: Basic Books.

Bellman, Beryl L. and Alex Tindimubona (1991) "Global Networks and International Communications: AFRINET," paper presented at the 34th Annual Meeting of the African Studies Association, St. Louis, Missouri, November 23-26.

Brohman, John (1996). *Popular Development: Rethinking the Theory & Practice of Development*. Malden, Massachusetts: Blackwell Publishers Ltd.

Browne, Robert S. (1984). "Africa's Economic Future: Development or Disintegration?" *World Policy Journal*, Volume 1, No. 4.

Business Day (2008) "Ugandan minister knocks South Africa's telecoms policy and High SAT3 prices." Retrieved June 30, 2008 from
http://www.balancingact-africa.com/news/back/balancing-act_400.html

Castells, Manuel (1996). *The rise of the network society*. Cambridge, Massachusetts: Blackwell Publishers, Inc.

—— (2003). *Power of Identity: The Information Age: Economy, Society and Culture*, Volume 2. Oxford, UK: Blackwell Publishers.

—— ed., (1985) *High technology, space, and society*. Beverly Hills, California: Sage Publications, Inc.

—— and Peter Hall (1994). *Technopoles of the world: The making of 21st century industrial complexes*. London and New York: Routledge.

Chiahemen, John (1999.) "Earthy image dogs Nigeria's Obasanjo," a Reuters wire story, Feb. 26, 1999

Chirot, Daniel (1994). *How societies change*. Thousand Oaks, California: Pine Forge Press.

Credé Andreas and Robin Mansell (1998). *Knowledge Societies ... in a nutshell: information technology for sustainable development*. Ottawa, Ontario: IDRC.

Dahrendorf, Rolf (1975). *The New Liberty*. London: Routledge & Kegan Paul.

Daily Champion (2000). "Telecoms expert, Nnama urges Nigeria to embrace 21 century IT."

Damkor, Moses et al (2001). *Report of the Steering Committee on Computerization of the Ministry of Communications*.

Dean, Edwin (1972). Plan Implementation in Nigeria: 1962-1966. Nigerian Institute of Social and Economic Research, Ibadan: Oxford University Press.

Dearnley, James and John Feather (2001). *The wired world: An introduction to the theory and practice of the information society*. London, England: Library Association Publishing.

Dordick, Herbert S. and Georgette Wang (1993). *The information society: A retrospective view*. Newbury Park, UK: Sage Publications.

Dos Santos, Theotonio (1998). "The Structure of Dependence" in Mitchell A. Seligson and John T. Passé-Smith (eds.), *Development and Underdevelopment: The Political Economy of Global Inequality*. Boulder, Colorado: Lynne Reiner, p.251-261.

Elendu, Jonathan, "Nigeria ... through my prism." Retrieved on Sept. 10, 2008 from http://elendureports.com/index.php?option=com_content&task=view&id=165&Itemid=29

Ellul, Jacques (1967). *The Technological Society*. New York: Vintage Books.

Emeagwali, Philip (1997). "Can Nigeria Vault into the Information Age?" A paper delivered at the 1997 World Igbo Congress in New York. Retrieved August 14, 2008 from http://emeagwali.com/speeches/igbo/Can-Nigeria-Leapfrog-into-the-Information-Age.pdf

Ette, Ezekiel (2008). "Happy birthday Nigeria." Retrieved on Oct. 1 from Annang-forum@yahoogroups.com

Federal Government of Nigeria (1999). Nigerian Economic Policy, 1999-2003. The Presidency, Abuja: National Orientation and Public Affairs.

Federal Ministry of Information (1989). 30 Questions and Answers on SAP ... and the Gains of SAP. Lagos, Nigeria.

Federal Republic of Nigeria (2001) National *Information Technology Policy*

—— (2000). *National Policy on Telecommunications*.

—— (1992). "Decree No. 75."

Feng, Tian (2008), president of Huawei in the Middle East and North Africa region, quoted in "Huawei's solutions support Africa's emerging economies." Retrieved Sept. 27, 2008 from http://www.ameinfo.com/156409.html

Forbes (2007). "UN'S ITU demands 'Marshall Plan' for Africa's internet connectivity." Retrieved Sept. 27, 2008 from
http://www.forbes.com/markets/feeds/afx/2007/07/11/afx3903509.html

Forrest, Tom (1995). Politics and Economic Development in Nigeria - Updated Edition. Boulder, Colorado: Westview Press, Inc.

Frank, André Gunder (1967). *Capitalism and Underdevelopment in Latin America: Historical Studies of Chile and Brazil*. New York: Monthly Review Press.

Friedman, Thomas L., (2001). *The Lexus and the Olive Tree*, 2nd edition. New York: Farrar, Straus & Giroux.

—— (2008). "Making America stupid." *The New York Times*, Sept. 13, 2008.

—— (2005). *The World is Flat*. New York: Farrar, Straus and Giroux.

Giddens, Anthony (1990). *The consequences of modernity*. Cambridge: Polity

—— (2000). *Runaway World: How globalization is reshaping our lives*. New York: Routledge.

Global Security Organization (2008). "Liberia - First Civil War - 1989-1996." Retrieved Sept. 15 from http://www.globalsecurity.org/military/world/war/liberia-1989.htm

Goodman, Seymour E., Grey E Burkhart, William A. Foster, Laurence I. Press, Zixiang (Alex) Tan and Jonathan Woodard (1998). *The Global Diffusion of the Internet Project: An Initial Inductive Study* (A Mosaic Group Report, March 1998).

Guardian Newspapers (2008). "Citizen Joseph Adumekwe and the Gabonese authorities." Editorial, August 10, 2008. Retrieved Sept. 23 from http://www.guardiannewsngr.com/editorial_opinion/article01//indexn2_html?pdate=10 0808&ptitle=Citizen

Hamelink, Cees J. (1997). *New information and communication technologies, social development and cultural change*, Geneva, Switzerland: United Nations Research Institute for Social Development Discussion Paper No.86.

Hammond, Allen L. (1998). *Which World? Scenarios for the 21st Century: Global Destinies, Regional Choices.* Washington, DC: Island Press.

Harding, Sandra (1987). *Feminism and Methodology.* Bloomington, Indiana: Milton Keynes; Indiana University Press.

Harvey, David (1989). *The condition of postmodernity: An enquiry into the origins of cultural change.* Oxford: Blackwell.

Heeks, Richard (2006). "Introduction: Theorizing ICT4D Research." Information Technologies and International Development. Vo. 3, Number 3, Spring 2006, 1-4.

—— (1999). *Information and communication technologies, poverty and development*, development informatics, Working Paper Series, No. 5. Manchester, U.K.: Institute for Development Policy and Management, University of Manchester.

Howkins, John and Robert Valantin (Eds.). (1997). *Development and the information age: Four global scenarios for the future of information and communication technology.* Ottawa, Canada: IDRC/UNCSTD.

Human Rights Watch (1997). "Vision 2010." Retrieved August 14, 2008 from http://www.hrw.org/reports/1997/nigeria/Nigeria-08.htm#P529_132846.

Ifodoe, David (2002) "President Obasanjo would do well by signing the Abrogation Bill into Law," in *Vanguard*, Friday, December 20, 2002. Accessed on December 27 at: http://www.vanguardngr.com/articles/2002/viewpoints/vp220122002.html

Ihonvbere, Julius O. (1994). *Nigeria: The Politics of Adjustment & Democracy.* New Brunswick, New Jersey: Transaction Publishers.

Ikpe, Okey (2005). "Microsoft trains FG officials, extends partnership with NITDA." Retrieved April 17, 2008 from http://www.microsoft.com/africa/press/ng_trains_fg.mspx

Inkeles, Alex and Smith, David H. (1963). *Becoming Modern: Individual Change in Six Developing Countries.* Cambridge, Massachusetts: Harvard University Press.

International Telecommunication Union (1998). *World Telecommunication Development Report: Universal Access.* Retrieved July 23, 2008 from http://www.itu.int/ITU-D/ict/publications/wtdr_98/WTDR98_e_chap2.pdf

—— (2008). ITU Report, June 2008. Retrieved July 23, 2008 from www.itu.int.org

Internet World Statistics (2008). "World Internet Usage and Population Statistics." Retrieved July 30, 2008 from http://internetworldstats.com/stats.htm

Jarmon, Charles (1988). *Nigeria: Reorganization and development since the mid-twentieth century.* Leiden, The Netherlands: E. J. Brill.

Keohane, Robert O. and Joseph S. Nye (1977). *Power and interdependence: World politics in transition.* Boston: Little, Brown.

Bibliography

Kendall, Kenneth E., ed. (1999). *Emerging Information Technologies.* Thousand Oaks, California: Sage Publications.

Leenes, Ronald and Bert-Jaap Koops (2005). "'Code': Privacy's Death or Saviour?" *International Review of Law Computer s& Technology*, Volume 19, No. 3, Pages 329–340.

Lerner, Daniel (1958). *The Passing of Traditional Society: Modernizing the Middle East.* Glencoe, IL: Free Press.

Lewis, W. Arthur (1967). "Reflections on Nigeria's Economic Growth." Paris, France: Development Centre of the Organization for Economic Co-operation and Development.

Light, Jennifer S., "The Digital Landscape: New Space for Women?" in *Gender Place & Culture: A Journal of Feminist Geography*, Sept.95, Vol. 2 Issue 2, p.134.

Lipton, Michael (1976). "Urban Bias and Inequality," excerpts from Why People Stay Poor: A Study of Urban Bias in World Development (1976) reprinted in Mitchell A. Seligson and John T Passé-Smith, eds. (1998) *Development and Underdevelopment: The Political Economy of Global Inequality*, 2nd edition. London, UK: Lynne Rienner Publishers, Inc.

Livraghi, Giancarlo (2008). "The Internet in Africa." Retrieved July 18, 2008 from http://www.gandalf.it/data/africeng.htm

Machlup, Fritz (1972). *The Production and Distribution of Knowledge in the United States.* Princeton, NJ: Princeton University Press.

Mackay, Hugh. (2000). "The globalization of culture?" in Held, David, ed., *A globalizing world? Culture, economics.* New York: Routledge.

Madamkor, Moses (2001). "Report on Investment Incentives in Nigeria to the Telecommunication Development Bureau, ITU-D Study Groups."

Maitland, Donald (1984). *The Missing Link: Report of the Independent Commission for Worldwide Telecommunications Development.* Geneva, ITU.

Makinde, Akin M. (1986). "Technology Transfer: An African Dilemma," in Murphy, John W., Algis Mickunas and Joseph J. Pilotta (eds.), *The Underside of High-Tech: Technology and the Deformation of Human Sensibilities.* New York: Greenwood Press.

Martinussen, John (1997). *Society, State and Market: A Guide to Competing Theories of Development.* Atlantic Highlands, New Jersey: Zed Books.

Masuda, Y (1981). *The Information Society as Post Industrial Society.* Washington, DC: The World Future Society.

Mehmet, Ozay (1999). *Westernizing the Third World: The Eurocentricity of economic development theories*, second edition. London and New York: Routledge.

Miller, Jonathan and Robert S. Day (2000). "Towards a national ICT policy and planning process: lessons from South Africa," in IFIP WG 9.4, *Conference Proceedings.* Conference 2000: Information Flows, Local Improvisations and Work Practices, Cape Town, South Africa, May24-26, 2000.

Montealegre, Ramiro (1999). "A case for more case study research in the implementation of Information Technology in less-developed countries." *Information Technology for Development*, Vol. 8, Issue 4

Morales-Gomez, Daniel and Martha Melesse (1998). "Utilizing information and communication technologies for development: The social dimensions," in *Information Technology for Development*, Vol. 8 Issue 1, p. 3-14.

Mowlana, Hamid (1997). *Global information and world communication*, Second edition. Thousand Oaks, California: Sage.

Mungai, Wainaina (2002). "The African Internet: Impact, winners and losers," A background paper prepared for the 2nd International conference of the African Youth Foundation (AYF) on Technology and Human Development in Africa, June 6-7, 2002, in Bonn, Germany.

Murphy, John W., Algis Mickunas and Joseph J. Pilotta eds. (1986). *The Underside of High-Tech: Technology and the Deformation of Human Sensibilities*. New York: Greenwood Press.

"Niger Delta Development Commission: Historical Background," available at http://www.nddconline.org/history.shtml

Nigerian Communications Commission (2008). "Industry Statistics." Retrieved July 23, 2008 from www.ncc.gov.ng

Nwankwo, Tony (2002). "Dichotomy Bill and the Conference Question," in *Vanguard*, Sunday, December 22, 2002. Retrieved on Dec.27, 2007 from http://www.vanguardngr.com/articles/2002/viewpoints/vp322122002.html

Obadare, Ebenezer (2001). "Constructing Pax Nigeriana? The Media and Conflict in Nigeria-Equatorial Guinea Relations." *Nordic Journal of African Studies* 10(1): 80-89.

Obasanjo, Olusegun (2000). "Nigeria on the Agenda: The Journey so far" – First anniversary speech.

Obijiofor, Levi (2001). "Singaporean and Nigerian journalists' perceptions of new technologies." *Australian Journalism Review*, Vol. 3: 1, p.131-151.

Okigbo, Pius N.C. (1989). *National Planning in Nigeria: 1900-1992*, Islington, London: James Currey Ltd.

Olaloku, F. A. (1979) *Structure of the Nigerian Economy*. New York, New York: St. Martin's Press.

Olufuye, Jimson (2001). "The state of IT practice in Nigeria" in *The Guardian*, July 24, 2001.

Omoigui, Nosa, "The Information Age as a Key to Nigeria's Renaissance: Opportunities, Risks and Geopolitical Implications," Available at: http://www.nigerianscholars.africanqueen.com/opinion/NosaTech.htm

Onishi, Norimitsu (1999). "Man in the news; Nigerian question mark: Olusegun Obasanjo." *The New York Times*, March 2, 1999.

Onyemelukwe, Clement C. (1966). *Problems of Industrial Planning and Management in Nigeria*. London, U.K.: Longmans, Green and Co. Limited; New York, New York: Columbia University Press.

Pew Internet & American Life Project (2007). "Demographics of Internet Users." Retrieved on August 18, 2008 from http://www.pewinternet.org/trends/User_Demo_6.15.07.htm.

Porat, Marc and M. Rubin (1977). *The information economy: Development and measurement*. Washington, DC: Government Printing Office.

Poster, Mark (1999). "National identities and communication technologies." *The Information Society*, 15.

Pye, Lucian W., ed. (1963). *Communications and Political Development*. Princeton, New Jersey: Princeton University Press.

—— and Sidney Verba, eds., (1965) *Political Culture and Political Development*. Princeton, New Jersey: Princeton University Press.

Ragin, Charles (1987). *The comparative method: Moving beyond qualitative and quantitative strategies*. Berkeley, California: University of California Press.

Rajaee, Farhang (2000). *Globalization on tTrial: The human condition and the information civilization*. Ottawa, Ontario: IDRC.

Bibliography

Rimmer, Douglas (1981). "Development in Nigeria: An Overview" in Bienen, Henry and V. P. Diejomaoh, Eds. The Political Economy of Income Distribution in Nigeria. New York; London, U.K.: Holmes & Meier Publishers, Inc.

Rostow, Walt W. (1961). *The Stages of Economic Growth: A Non-communist Manifesto*. Cambridge, U.K.: Cambridge University Press.

Schiller, Herbert I. (1986). *Information and the crisis economy*. New York: Oxford University Press.

—— (1984). "New Information technologies and old objectives," *Science and Public Policy*.

Seligson, Mitchell, A (1998). "The Dual Gaps: An Updated Overview of Theory and Research" in Mitchell A. Seligson and John T. Passé-Smith (eds.), *Development and Underdevelopment: The Political Economy of Global Inequality*. Boulder, Colorado: Lynne Reiner, p.3-8

—— and John T. Passé-Smith, eds., (1998). *Development and Underdevelopment: The Political Economy of Global Inequality*. Boulder, Colorado: Lynne Reiner.

So, Alvin (1990). Social Change and Development: Modernization, Dependency and World-System Theories. Newbury Park, California: Sage Publications Ltd.

Spender, Dale "The Position of Women in Information Technology – or Who Got There First and with What Consequences?" in *Current Sociology* April 1997, Vol. 45(2)

Stover, William James (1984). *Information Technology in the Third World: Can I.T. Lead to Humane National Development?* Boulder, Colorado: Westview Press/A Westview Replica Edition.

Stuff (2008). "Man loses $20k in online love scam." Retrieved Sept. 29 from http://www.stuff.co.nz/4708739a28.html

Sutherland, Ewan (2008). "Counting mobile phones, SIM cards & customers." The Link Center. Retrieved Sept. 27 from http://www.itu.int/ITU-D/ict/statistics/material/sutherland-mobile-numbers.pdf

The Economist (2007). "The digital gap: More than a click to put Africa online." Retrieved on Sept. 29, 2008 from http://www.economist.com/world/mideast-africa/displaystory.cfm?story_id=9990626

The Washington Times (1999). "My brother's keeper: Leading regional peacekeeping efforts" – An international report prepared by the *Washington Times* Advertisement Department, September 30, 1999.

Tucker, Vincent (1999). "The Myth of Development: A Critique of a Eurocentric Discourse" in Munck, Ronaldo and Denis O'Hearn, eds. *Critical Development Theory: Contributions to a New Paradigm*. London and New York: Zed Books.

UNESCO (1955). *Research paper no. 9*.

United Nations Development Program, *Human Development Report*, 2001 and 1999

United Nations Economic Commission for Africa (2002). *Economic Report on Africa 2002: Tracking Performance and Progress*.

Usoro, Paul (2000). "Matters arising on mobile wireless licensing," *The Guardian*, October 31, 2000, p.43

Uwadibie, Nwafejoku Okolie (2000). *Decentralization and Economic Development in Nigeria: Agricultural Policies and Implementation*. Lanham, Maryland: University Press of America, Inc.

Van Reijswoud, Victor (2006). "Strategies for increased internet growth in Africa." Retrieved Sept. 29 from http://www.regulateonline.org/content/view/861/76/

Van Zoonen, Liesbet, "Feminist Theory and Gender," in *Media, Culture and Society* (SAGE, London, Newbury Park and New Delhi) Vol. 14 (1992)

Weatherby, Joseph, et al. (2006). *The Other world: Issues and politics of the developing world.* Seventh Edition. Longman.

Webster, Frank (1995). *Theories of the information society.* London: Routledge.

—— and Kevin Robins (1986). *Information Technology: A Luddite Analysis.* Norwood, New Jersey: Ablex Publishing Corporation.

Williams, Gavin, ed. (1976). Nigeria: Economy and Society. London, U.K.: Rex Collings Ltd.

Wolcott, Peter, Seymour E. Goodman and Grey E. Burkhart (1997). *The Information Capability of Nations: A Framework for Analysis* (A Mosaic Group Report January 1997).

World Bank (2008). "Data and Statistics." Retrieved Sept. 16, 2008 from http://www.worldbank.org

—— (2002) *World Development Indicators,* April 2002

World Economic Forum (2002) *The Global Competitiveness Report, 2001-2002.* Oxford University Press.

Wright, Stephen (1998). *Nigeria: Struggle for Stability and Status.* Westview Press.

Yesufu, Tijani M. (1996) *The Nigerian Economy: Growth Without Development.* University of Benin, Benin City, Nigeria: The Benin Social Science Series for Africa.

Zartman, I. William, with Sayre Schatz (1983). "Introduction" in I. William Zartman, ed., *The Political Economy of Nigeria.* New York, New York: Praeger.

Index

• A •

Abacha, 62, 63, 64, 67
Abuja Technology Village, 180
   Computer villages, 180
Accord International, 171
Adebo, 52
ADSL, 165
African Development Bank (ADB), 22
African Renewal, 171
African Union (AU), 169
AfriSPA, 181
Ahemba, 167, 168
Ajayi, 9
Aka, 61
Akpan-Obong, 184
   and Parmentier, 34, 178
Almond and Verba, 19
Alternative sources of power, 164
Aluko, 167, 168
Amaefule, 102
Amana, E,
Andre Gunder Frank, 20
Anunobi, F, 56
Aragba-Akpore, 76, 79
Aron, 176
Aremu, 66
ARPANET, 6, 165
Arnold, 127
Asika, 171
Aso Rock, 100, 101
Association of Telecommunications
   Operators of Nigeria, 89
Avgerou, 5, 25
Awareness and Access, 88
   access, 152, 181

• B •

Babangida Administration, 56, 59, 62, 66
Bada, 26
Band-wagon approach, 27
Basic-needs Strategy, 21
   see development theory, 17
Bates, 54
Bell, 1, 28, 29
Big Brother, 168, 170, 172, 178
Bretton Woods, 18
   IMF, 18, 56, 58, 60, 67
   World Bank, 1, 57, 60, 159, 165, 167
Broadband,
Brohman, 21, 23, 25
Browne, 22
Buhari government, 58
Bush, 166
Businessday, 177

• C •

Capital flight, 46
Cargo cult, 34, 35, 36
Casablanca Bloc, 169
Case study approach, 11
Castells, 1, 6, 7, 20, 28, 30, 31, 34
   Cell phone world, 164
Celtel (Zain), 80, 81, 82, 160, 161, 175
Central Bank of Nigeria, 57
Chiahemen, 66
Chirot, 8
Code Division Multiple Access (CDMA), 79
Colonial Development, 48
Communication revolution, 7

Computer Association of Nigeria (COAN), 84, 89
Computer Professions Registration Council of Nigeria, 89
computer smart, 117
Consumer Affairs Bureau, 174
Cooperative Information Network (COPINET), 84
Core development strategy, 1
Corps member, 126, youth corpers, 126, 142, 153
Crede and Mansell, 4, 5
Critical Theory, 33
Culture elites, 8
Culture of consumerism and waste, 45
Cyber café, 79, 127, 132, 135, 143, 152, 153, 184
    Business centers, 136, 140, 141, computer aceess, 140, 181

• D •

Dahrendorf, 28, 29
Damkor, 78, 106
Data services,173
Dean, 50, 52
Dearnley and Feather, 5, 6, 7, 29, 30
Decree No.75 of 1992, 66
Democratization of information, 167
Dependency Theory, 20
  colonial, 20
  financial, 20
  technological-industrial, 20
Detours, 12, 156
    infrastructure detours, 163
    ideological and cultural detours, 169
      ethnicity, 170
Development Information, 25
Development Theory, 17
    basic needs strategy, 21
    dependency theory, 20
    modernization theory 19, 20, 25
DFRRI, 61
Digital Mobile Licenses (DMLs), 73, 79, 82, 83, 98, 147, 164, 173
 Digital mobile phone market, 73

Dordick and Wang, 18, 19, 21, 23, 24, 25, 27, 28, 29, 32
Dos Santos, 20
DPRS, 106, 107, 108, 109, 110
Dystopian/continuity approach, 28
Dystopian/pessimistic perspectives, 27

• E •

ECOMOG, 170
Econet (Celtel, Zain), 80, 81, 82, 160, 161, 175
Economist, The, 182
ECOWAS, 22, 170
EEC, 169
EFCC, 184, 185
Ekuwem, 81, 84, 87, 162
Electronic commmerce, 157
Elendu, 167
Ellul, 28, 29, 30
Emeagwali, 64
Eminent Persons Group, 169
e-revolution, 118
Ette, 184
Export Free Zone, 46
Export Processing Zones, 86
Exposure Theory of Communication, 23
External-oriented Communication, 139

• F •

Federal character principle, 170
Federal government of Nigeria, 65, 75, 76,77, 80, 84, 87
Feng, 178
Five indicators of development, 26
  literacy , 26
  health, 26
    income and economic welfare, 26
    choice, democracy and participation, 26
  technology, 26
Footloose Multinational Corporations (MNCs), 46
Forbes, 183
Forecast Planning, 47

Index

Forrest, 60
Fortress approach, 27
French System of Indicative Planning, 47
Friedman, 31, 166, 167, 178

• G •

GDI framework, 114, 115, 116, 144
General Yakubu Gowon, 57, 126, 170
Giant of Africa, 167, 172, 188
Giddens, 4
Globacom, 83
Global Diffusion of the Internet (GDI), 98, 114, 115, 116, 119
Global hegemon, 3
Global Information Society, 88, 108, 156, 163, 167
Global Network Society, 6, 7, 30, 172, 178
Global Security Organisation, 170
Global system of mobile communication (GSM), 10, 79
Global welfarism, 49, 67
Goelro Plan of 1920, 47
Goodman, et al, 114, 116, 143
Guardian Newspaper, The, 168

• H •

Habermas, 4
Haggard and Kaufman, 56
Hamelink, 4, 27, 28
Hammond, 34
Harvey, 7
Heeks, 5, 7, 26, 27, 37
Held, 7, 31
Herbert, 170
Howkins and Valantin, 5, 26, 33, 34, 35, 36, 37, 137
Hugh, 7
Human Rights Watch, 64
Huawei Technology, 177, 178

• I •

ICTs, 1, 2
  landscape,
  sector
ICT diffusion, 2
ICT Indicators, 175
ICT4D discourse, 7, 26, 37, 119, 137, 181, 186
Ifidon, 169
Ihonvbere, 46, 56, 59
IMF conditionalities, 60
IMF debates, 59
Impact causes, 27
Industrial-age, 118, 119
Information age, 6
Information Cyber, 104
Information networks, 30, 31
Information processing technologies, 5
Information revolution, 6, 7
Information society, 6, 30, 31, 32, 166, 167, 172
Information superhighway, 9, 12, 173
Information Technology (IT), 5
Information Technology Association of Nigeria (ITAN), 84, 89
Informational mode of production, 1, 5, 177
Infrastructure sharing, 82
Inkeles and Smith, 24
Insecurity of life and property, 46
Institute of Software Practitioners of Nigeria (ISPN), 84, 89
Interconection, 82, 83
Interconectivity, 174
  agreement, 83
International development agencies, 1
  Canadian International Development Research Center (IDRC), 1
  Knowledge for Development Program, 1
  G8 (Digital Opportunities Taskforce), 1
  United Nations Millenium Development Goals (MDGs), 1, 2, 176
International Telecommunications Union, (ITU) 4, 8, 127, 128, 150, 153, 165, 166, 175, 182
International Telegraphs and Telecommunications (ITT), 3

Internet-on-demand, 113
Internet Service Provider (ISP), 152, 174, 183
   Subscription, 128, 164, 175
      cell phone, 129, 133, 135, 136, 175
      internet, 128, 129, 132
Internet world statistics, 6, 165
IT free zones, 86
IT person, 167
IT project unit, 104
IT Solutions, 89, 97, 98
IXP, 183

• J •

Jarmon, 10, 49, 52

• K •

Keohane and Nye, 1, 6, 23
Kendall, 33, 37, 114
Kendall framework, 114

• L •

Lagos Plan of Action (LPA), 22, 171
Land and labor, 43
Leenes and Koops, 125
Legal framework, 157
Lerner, 23, 24, 45
Lewis, 45
Liberalization of imports, 53
Licenses, 79, 80, 173, 174
   import license, 57
   digital mobile licenses, 73, 79, 173
   unified licenses, 79, 81, 82, 143, 173
Light, 165
Lipton, 54
Local area network (LAN), 87, 107, 112
Local service providers, 73
Log-in point, 135

• M •

Machclups, 29, 30
Maitland, 4
Makinde, 34
MAMSER, 167
March of follies, 34, 35
Marshall Plan, 183
Martinussen, J, 17, 23, 25, 34
Masuda, 29
Marxism, 47
m-banking, 118, 176
Mehmet, 19
Miller and Day, 34
Ministry of Information and
   Communications, 99, 103, 162
      Ministry of Communications, 106, 108, 115
      Ministry of Information and National Orientation, 99, 103
Ministry of Education, Federal, 99, 110, 112, 117
Ministry of Information, Federal, 60, 99
Ministry of Science and Technology, 74, 99, 100, 102
Mobile commerce, 157
Mobile Phones, 173, 174, 175, 187
   Cellular, 175
Mobile telephony, 79, 127, 147, 164, 174, 175, 187
   GSM, 79, 80, 81, 82
   CDMA, 79, 81, 82
Modernization Theory, 19, 20, 25
Modernizing elites, 8
Monrovia, 169
   Bloc, 168
   Doctrine, 170
Montealegre, 11, 156
Morales-Gomez and Meleese, 32
Mowlana, 7, 24
MTN, 80, 81, 82, 175
Muchie, 181
Munck and O'Hearn, 17
Mungai, 26, 27
Murphy, et al, 125

Index

• N •

National Carriers, 83
National Committee for the Acquisition of Computer and Electronic Technology (NACACET), 74
National Economic Planning Commission, 62
National Economic Recovery Plan, 60
National Policy on Education, 88, 90
National Policy on Information Technology (NPIT), 75, 83, 84, 85, 87, 88, 156
National Policy on Telecommunications (NPT), 73, 75, 76, 84, 99, 156
NEPA, 161
NEPAD, 171
Netblocs, 34, 35
Network age, 5
Network society, 30, 31
Networld, 34, 36
New Information and Communications Order, 4
Nigerian Communications Commission (NCC), 66, 73, 74, 75, 76, 78, 79, 80, 83, 99, 112, 113, 115, 116, 127, , 164, 172, 173, 174
Nigerian Information Technology Development Agency (NITDA), 84, 85, 87, 101, 102, 112, 114, 115, 117, 118 ACT, 172, 173
Nigerian Internet Group, 81, 84, 91
Nigerian Planning Commission, 66
   Omotosho, A, 66
Nigerian Telecommunication Limited (Nitel), 80, 76, 160, 172
NITDEF, 85
Nollywood, 179, 186

• O •

OAU, 21, 168, 169, 186
Obadare, 168
Obasanjo, O, 57, 64, 65, 67, 118, 147, 159
Ocean transport, 47
Official secrets, 166
Ogoni Nine, 170
Okigbo, 47, 48, 49, 50, 51, 52, 53, 54, 55, 57, 58, 62
Olaloku, 44
Olufuye, 74, 75
Omo-Ettu, 63, 83, 152, 173
Omoigui, 170
Onishi, 169
Onyemelukwe, 52
Oyawoye, 100, 102, 167, 168

• P •

Paradigms of progress, 28
   the new liberty, 28
   post-industrial society, 28
   technological society, 28
Pervasiveness, 99, 114
Petro-dollar, 53
Pew Internet and American Life Project, 126
Political instability, 46
Porat and Rubin, 29, 30
Poster, 6, 7
Post-fifth National Development Plan, 62
Post Independence National Development Plan, 51
Potholes, 12, 156
   the institutional framework, 156
      policy implementation, 157
      legal framework, 157
   state of the infrastructure, 160
      telecommunications infrastructure, 160
      constraints of electricity, 161
      import dependency, 162
   levels of poverty and illiteracy, 165
   cultural framework and ethnicity, 165
Precondition for takeoff, 45
Predatory practice
Presidency, The, 99, 102
Private Telecommunications Operator (PTOs), 76, 80, 81, 82, 98, 133, 172
Processes of globalization, 1
Providers, 79, 80, 83, 152
PSnet, 173
Public switched telecommunication network (PSTN), 81
Pye, 24

• Q •

Queensland Police Service, 180

• R •

Ragin, 11
Rail transport, 47
Rajaee, 5, 6
Rimmer, 49, 50
Rostow, 19, 29, 44, 45

• S •

Safaricom Kenya, 176
SAP, 56, 60, 67
SAP conditionalities, 60
SAT 3, 183
Satellite communication, 79
Scenario-modeling, 34, 37
Schiller, 4, 5
Second National Development Plan, 51
Seligson, 25
Sectoral absorption, 99, 116, 117
Shagari administration, 57, 58
Shonekan, 62
Six Years of Telecom Revolution, 98
Sleeping Giant, 167, 168, 187
So, 25
Sophistication of use, 99, 115
Spatio-temporal compression, 7
Spender, 165
Stage approach, 33, 37
State creation, 45
Stover, 24
Stuff, 184
Structuralism, 33
Suitcase incident, 58
Sutherland, 176
Switching facilities,

• T •

Tariff, 73, 74, 80, 81, 187
Techno-centric utopian, 27
Technological transfer, 34
Technology impacts, 27
Telecom Act, 172
Teledensity, 127, 150, 151, 152, 174
Telemedicine, 26
Ten-year Plan of Development and Welfare
 for Nigeria, 49, 50, 67
The Babangida years, 59
The Buhari administration, 58
The Fifth National Development Plan, 60
The Fourth National Development Plan, 55
The Structural Adjustment Program years, 56
The Third National Development Plan, 53
Top-down Development Model, 21
Transport network, 47
 ocean transport, 47
 rail transport, 47
 marine, 66
Tucker, 18

• U •

UiXIP, 183
Umbrella people, 126, 127, 152, 176
 Money transfer agents, 176
Umesao, 29
UNECA, 166
UNESCO, 3
USE IT, 147
Usoro, P, 78
Uwadibie, N, 45

• V •

Van Rejiswoud, 181
Video conferencing,
Virtual road, 155, 156
Vision 2010, 63, 64
VOIP, 82, 172

VSATs, 85, 87, 104, 114, 116, 117, 132, 160, 172, 173, 174

• W •

War Against Indiscipline, 58, 59
Washington Times, The, 170
Weatherby, 9
Webster, 6, 30, 31, 32, 33
Welfarism, 49
Wide Area Network (WAN), 100
Williams, 48, 52
Wireless access,127, 140
   Hot spots
   fixed lines, 73, 81, 127, 138, 150, 160,164, 174
   wireless fixed lines, 81, 127, 128, 139, 160, 164, 173, 187
Wireless Local Loops (WLL), 76
Wolcott, 37, 98, 114, 115, 128
World Bank standards, 44
World Factbook, 159
World telecommunication, 175
Wright, 62, 63

• Y •

Yahoo boys, 146, 158, 185, 188
   419 scam, 157, 158, 183, 184
Yar'Adua, 120
    government, 67
Yesufu, 43, 44, 45, 46, 47, 48, 49, 53, 55, 56, 59, 60, 62, 163

• Z •

Zain (Celtel, Econet), 80, 81, 82, 160, 161,175
Zartman, 44, 57
Zinox Technologies, 90, 102, 179, 180
Zoonen, 166, 169

SOCIETY AND POLITICS IN AFRICA

Yakubu Saaka, General Editor

This multidisciplinary series publishes monographs and edited volumes that provide innovative approaches to the study and appreciation of contemporary African society. Although we focus mainly on subjects in the social sciences, we will consider manuscripts in the humanities that treat context as a significant aspect of discourse. Within the social sciences, we are looking for not only analytically outstanding studies but, what is more important, ones that may also have significant implications for the formulation and implementation of public policy in Africa. We are especially interested in works that challenge pre-existing hierarchies and paradigms.

For additional information about this series or for the submission of manuscripts, please contact:

>Peter Lang Publishing
>Acquisitions Department
>29 Broadway, 18th Floor
>New York, New York 10006

To order other books in this series, please contact our Customer Service Department:

>800-770-LANG (within the U.S.)
>(212) 647-7706 (outside the U.S.)
>(212) 647-7707 FAX

Or browse online by series at:

>www.peterlang.com

www.ingramcontent.com/pod-product-compliance
Ingram Content Group UK Ltd.
Pitfield, Milton Keynes, MK11 3LW, UK
UKHW021833140426
5217IPUK00021B/1427